Objective Environmental Pollution and Climate Change

Related Books

Titles	*Authors*
— A Text Book of Environmental Science	*Vidya Thakur*
— Agricultural Input and Environment	*S.P. Palaniappan*
— Alkali Wastelands Environment and Reclamation	*I.C. Gupta*
— Applied Microbiology (Agriculture, Environmental, Food and Industrial Microbiology)	*S.M. Reddy*
— Biodiversity and Environmental Biotechnology	*P. Dwivedi*
— Climate Change and Environment	*J. Sundaresan*
— Climate, Agriculture and Man Shrinking Biodiversity and Sustainability	*V.K. Sehgal*
— Disaster Perception and Management	*T.S. Chouhan*
— Ecology of Desert Environments	*I. Prakash*
— Environment at Crossroads Challenges and Green Solutions	*Arun Arya*
— Environment Impact Assessment for Wetland Protection	*V.S. Kulkarni*
— Environmental Design and Development	*H.S. Sharma*
— Environmental Engineering and Safety	*Sangeeta Raut*
— Environmental Management in Mining Areas	*N.C. Saxena*
— Ecology Workbook	*R. Mishra*
— Environmental Pollution and Environmental Management	*P. Dwivedi*
— Environmental Protection	*A.K. Thukral*
— Environmental Resources and Tribes of North Western India	*V. Sankaran*
— Essentials of Environmental Studies	*S.S. Sagwal*
— Geo-informatics For Combating Land Degradation and Desertification	*T.S. Chouhan*
— Human Impact on Desert Environment	*P. Narain*
— Hydrogeology and Environmental Geology of Sonar River Basin	*S.K. Jain*
— Industrial Pollution and Environmental Management	*R.K. Trivedy*
— Industrial Waste Waters and Environmental Pollution	*D.C. Joshi*
— Mining Environment Management Manual	*N.C. Saxena*
— Objective Environmental Studies and Disaster Mgmt.	*M. Prasanthrajan*
— Pollution Risk and Impact on Social, Economic and Environmental Sectors and Remediation	*Golam Kibria*
— Prayavaran Javiki: Environmental Biology	*M.M. Trigunayat*
— Text Book of Disaster Management A.K. Shrivastava	
— Questions and Answers in Environmental Science Practical	*R.K. Kamble*
— Wasteland Management and Environment	*A.K. Roy*
— Wildlife Ecology And Conservation (21st Century Biology And Agriculture: Textbook Series)	*M. Balakrishnan*

OBJECTIVE ENVIRONMENTAL POLLUTION AND CLIMATE CHANGE

M. Prasanthrajan

Associate Professor (Environmental Science)

Tamil Nadu Agricultural University,
Coimbatore, Tamil Nadu
India

SCIENTIFIC PUBLISHERS

Published by:

SCIENTIFIC PUBLISHERS (INDIA)
5A, New Pali Road, P.O. Box 91
Jodhpur 342 001, India
E-mail: info@scientificpub.com
Website: www.scientificpub.com

ISBN: 978-93-90495-33-7
eISBN: 978-93-90495-34-4

Printed in India

PREFACE

This book is designed to meet the needs of students who enroll environmental science and those aspiring for competitive examinations. A vast literature on this subject is available but this book is a single volume discussion covering all the aspects of environmental pollution and climate change. This book is simple to read as it is designed with Multiple Choice Questions. Much effort has taken to provide all possible information and the subject is discussed in multiple angles to cater the diverse interests of the reader. This book is divided into twenty chapters covering Environmental pollution, Air Pollution, Atmospheric chemistry, Soil pollution, Heavy metal pollution, Water pollution, Thermal and radioactive pollution, Noise pollution, Plastic pollution, agricultural pollution, light pollution, hazardous waste and e-waste, Pollution control, Solid waste management, Climate change, Important environmental days, World environment day theme and host country, World water day themes, Air quality standards & Air quality index (AQI), Water quality standards, Noise quality standards, Important Climate Facts and COPs. Needless to say this book would be highly useful to the under graduate and post graduate students of science subject. University faculty dealing with environmental pollution and climate change would find this book as valuable reference book. In fact this book is not only meant for students and teachers but also for the people who are very much concerned with the well being of our environment. Though it is not possible to include all the information for each chapter and every term associated with environmental pollution and climate change but effort has been made to include all the information relevant to pollution and climate change. Since, most of the students use MCQ as one of their primary concerns, concentrated effort has been made to write a book that is highly usable to understand the basic and fundamental concepts of Environmental pollution and Climate change.

M. PRASANTHRAJAN

ACKNOWLEDGEMENT

I thank the Scientific Publishers and their editorial team in bringing this book to a finest shape. I thank Dr. K.T. Parthiban, Ph.D. the Dean (Forestry), Forest College and Research Institute, Mettupalayam for his excellent guidance in writing books and his vision in bringing the institute to a greater height. I thank Dr. A. Balasubramanian, Professor and Head (SNRM), and my students Sushant, Kirti, Manivenkatesh, Askari, Divakar, Krishnamoorthi, Mounika, Bobby for helping me in writing this book. Also, I thank my wife (Dr. Shalini) and son (Rishant) for their constant encouragement to write this book during the lock down period. I thank the University authority for their constant support for my academic career. Despite all, I thank my parents and Lord Almighty for giving me an opportunity to enrich the knowledge of the students through this book.

M. PRASANTHRAJAN

CONTENT

CHAPTER 1

ENVIRONMENTAL POLLUTION

1. **When is World Earth Day celebrated?**
   (A) June 5 (B) April 22
   (C) September 16 (D) October 16
2. **The satellite used to measure the ozone thickness of Antarctica region ________**
   (A) Nimbus 7 (B) Nimbus 6
   (C) Nimbus 5 (D) Nimbus 4
3. **What is the total mass of atmosphere?**
   (A) 5.7×10^5 tons of air (B) 5.7×10^{10} tons of air
   (C) 5.7×10^{12} tons of air (D) 5.7×10^{15} tons of air
4. **What is the permitted limit for SO_2 in the residential and rural area?**
   (A) 30 ug/m^3 (B) 120 ug/m^3
   (C) 80 ug/m^3 (D) 50 ug/m^3
5. **Which among the following is not a variable component of the atmo-sphere?**
   (A) Argon (B) Carbon dioxide
   (C) Methane (D) Ozone
6. **What is the maximum allowable noise level at residential area during night time?**
   (A) 55 dB (B) 45 dB
   (C) 60 db (D) 70 dB
7. **Which pollutant is generated when blasting is used in underground mines?**
   (A) SO_2 (B) NOx
   (C) CO (D) HF
8. **Which of the following is a bioreme-diation technique?**
   (A) Purging of air (B) Composting
   (C) Oxidation (D) Reduction

9. **Which among the following is not a thermoplastic?**
   (A) Polyethylene (B) PVC
   (C) Polypropylene (D) Polyester

10. **Plastics with thickness of ______ cannot be used for recycling.**
    (A) <30 microns (B) <50microns
    (C) <20 Microns (D) <60 microns

11. **When was the Water (Prevention and control pollution) Act passed?**
    (A) 1972 (B) 1974 (C) 1986 (D) 1982

12. **The Environmental Protection Act was enacted in the year ______________**
    (A) 1972 (B) 1976 (C) 1986 (D) 1982

13. **The First UN conference on human environment was held in ______________**
    (A) Rio de Janeiro (B) Kyoto
    (C) Stockholm (D) Montreal

14. **The Earth Summit of 1992 was held in ________**
    (A) Sweden (B) Rio, Brazil
    (C) Denmark (D) Norway

15. **Which of the following compound is responsible for the reduction of oxygen carrying capacity of the blood?**
    (A) CO_2 (B) CO
    (C) NO (D) NO_2

16. **Which among the following industry does not release Mercaptans?**
    (A) Pulp and paper manufac-turing
    (B) Tanneries (C) Refineries
    (D) Coal Tar

17. **At which point is the maximum dust concentration observed in a cement industry?**
    (A) Limestone crushing
    (B) At the kiln firing end
    (C) Coal crushing point
    (D) Loading of cement in wagons

18. **What is the main reason behind the production of more dioxins during the incineration of biomedical waste than the incineration of municipal solid waste?**

(A) Human tissues
(B) Chlorine and chlorinated products
(C) High moisture content
(D) Catalytic compounds

19. **Which pollutant is released during production of Aluminum from Alumina?**
(A) SO_2 (B) NO_2
(C) HF (D) CO

20. **Which among the following toxic gas is released during the production of phosphatic ferti-lizers from rock phosphates?**
(A) HCL (B) Phosphene
(C) HF (D) SO_2

21. **The section6 (2)(B) of Environment (Protection) Act 1986 is related to ______**
(A) Disposal of hazardous wastes
(B) Handling of hazardous wastes
(C) Maximum allowable limit of environmental pollutants
(D) Penalties for violating the Environment Act

22. **Which part of the Indian constitution states the following – "The state shall endeavor to protect and improve the environment and to safeguard the forests and wildlife of the country"?**
(A) Article 42 A (B) Article 51 A
(C) Article 56 A (D) Article 48 A

23. **Unburned carbon particles known as**
(A) Dust (B) Grit
(C) Soot (D) Smoke

24. **Which among the following is a toxic form of chromium?**
(A) Cr(IV) (B) Cr(II)
(C) Cr(III) (D) Cr(VI)

25. **The Chernobyl Nuclear disaster occurred in the year ______**
(A) 1966 (B) 1976
(C) 1986 (D) 1996

26. **What is the maximum fine a polluting industry can pay according to the Environment (Protection) Act 1986?**
(A) 15 Lakh (B) 50 thousand
(C) 1 Lakh (D) 5 Lakh

27. Which metal poisoning causes Itai – Itai disease?
    (A) Hg (B) Ni
    (C) Pb (D) Cd

28. Which heavy metal present in Antidandruff shampoo?
    (A) Hg (B) Se
    (C) Ni (D) Cr

29. The head quarters of UNEP is in ______
    (A) Nairobi (B) Brussels
    (C) New York (D) London

30. Ozone thickness is measured in ______
    (A) Curie (B) Cutie-Pie
    (C) Dobson (D) Decibel

31. Which one of the following were responsible for matters related with air pollution before enactment of the air (Prevention and Control of Pollution) Act, 1981?
    (A) Central Cabinet Ministers
    (B) CPCD (Central Pollution Control Board)
    (C) Central Board for Prevention and Control of Water pollution
    (D) Central Public Works Department

32. Which of the following laws have the Provision to establish or recognize laboratories for analysis of samples?
    I. The water (Prevention and Control of Pollution) Act, 1974
    II. The air(Prevention and Control of Pollution) Act 1981
    III. The environment(Protection) Act 1986
    (A) All of the three (B) Only (I)
    (C) Both (I) and (II) (D) Only (III)

33. Materials and substances that are likely to cause environmental pollution during the manufacturing process should be examined under the ______
    (A) The Factories Act (B) The Water Act
    (C) The Environment Act (D) The Air Act

34. If a person applies for consent to discharge industrial effluents, but receives no response from the state pollution Control Board then, how long should he wait to assume the 'unconditional consent'?
    (A) 2 months (B) 4 months
    (C) 6 months (D) 3 months

35. **Which act mentions the procedures and safeguards for handling of hazardous substances?**
    (A) The Factories Act, 1948
    (B) The water Act, 1974
    (C) The Environment Act, 1986
    (D) The Air Act, 1981

36. **The notice period to government for cases against violation of air Act, 1981 by an individual is ________**
    (A) 15 days (B) 60 days
    (C) 90 days (D) 120 days

37. **What is the significance of Air (Prevention and Control of Pollution) Amendment Act, 1987?**
    (A) Establishment of central pollution control Board
    (B) Inclusion of noise in definition of air pollution
    (C) Increase of noise in definition of air pollution
    (D) Public interest litigations by Public against polluting units

38. **When was the Central Board for Prevention and Control of water Pollution (CBPCWP) constituted?**
    (A) 1964 (B) 1970
    (C) 1974 (D) 1980

39. **Which of the following was the idea behind the water (Prevention and Control of Pollution) Cess Act 1977?**
    (A) Generate awareness about water pollution in industry
    (B) Collect revenues by imposing cess on water used
    (C) Reduce the use of water in industry
    (D) Collect revenue by imposing cess on effluent discharges

40. **For how many days is the notice served to a polluting industry under section 49 of the water (Prevention and control of Pollution) Act 1974?**
    (A) 30 days (B) 45 days
    (C) 60 days (D) 15 days

41. **Under the Water (Prevention and Control of Pollution) cess Act, 1977. A cess has to be paid for water consumed by ________ ________**
    (A) All industries
    (B) Certain industries only
    (C) Certain industries and local authorities
    (D) Local authorities and all Industries

42. On December 20, 1986 in a famous verdict Justice P. N. Bhagwati of Supreme Court laid down the principle that the compensation should be related to the magnitude and capacity of the enterprise so that it has a deterrent effect. It was related to which case?
(A) Bhopal gas tragedy case (B) Shriram oleum leak case
(C) Century Chlorine leak case
(D) NFL, Panipat, ammonia leak

43. Which one is considered as the biggest obstacle in managing municipal solid wastes in India?
(A) Collection (B) Segregation
(C) Transportation (D) Disposal

44. Which compound is present in normal writing paper?
(A) Cellulose only
(B) Cellulose and Hemicelluloses
(C) Lignin and Cellulose
(D) Hemicelluloses and lignin

45. Which pollutant emitted by the municipal solid waste incinerators causes birth defects, infertility and cancer even in a minute?
(A) Mercaptans (B) Dioxins
(C) Carbamates (D) Vinyl chloride

46. Which pollutant can mimic hormones and also known as environmental estrogen?
(A) Carbamates (B) Dioxins
(C) Vinyl chloride (D) Mercaptans

47. At what temperatures should furnace and incinerators be operated to reduce the formation of dioxins?
(A) 500 °C (B) 1200 °C
(C) 750 °C (D) 950 °C

48. Which act has the detailed description of the powers and duties of Chairman and Member secretary of central pollution Control Board?
(A) Water (Prevention and Control of Pollution) Act 1974
(B) Water (Prevention and Control of pollution) Rules 1975
(C) Environment (Protection) Act, 1986
(D) Air (Prevention and Control of Pollution) Amendment Act 1987

49. **Which of the following is not required for getting consent under section 25 or section 26 of the water (Prevention and Control of Pollution) Act when establishing an industry?**
(A) Raw material and chemicals to be used per month
(B) Quantity of solid waste generation expected
(C) Qualification of person in charge of the Effluent Treat-ment Plant
(D) Total number of employees expected to be employed

50. **Where can a person appeal when aggrieved by an order made by the state pollution Control Board under section 25, 26 or 27?**
(A) Central Board
(B) State board
(C) Appellate authority supreme court
(D) Supreme court

52. **How long can the members of a state pollution control board except the member secretary hold the office?**
(A) Six years (B) Four years
(C) Three years (D) Five years

51. **How is the Chairman of state pollution Control Board selected?**
(A) Nomination by central Govt.
(B) Nomination by state Govt.
(C) Election by the members of state board
(D) Nomination by members secretary of state board

53. **How is the Chairman selected in a joint pollution control board involving a Union territory?**
(A) Nomination by the government of the union territory
(B) Nomination by the Central Government
(C) Nomination by the state Government
(D) Election by the nominated members

54. **Byssinosis, also called "brown lung disease" or "Monday fever", is an occupational lung disease caused by exposure to**
(A) Cotton dust (B) ZnO
(C) CO (D) CO_2

55. **Which among the following industrial effluent will have more of BOD?**
(A) Paper (B) Dye
(C) Sugar (D) Rubber

56. Which instrument used to measure heavy metals?
(A) GC (B) AAS
(C) Colorimeter (D) HPLC

57. Aqua regia is a mixture of ________
(A) 3:1 HNO_3 HCl (B) 3:1 H_2SO_4 : HNO_3
(C) 3:1 HCl :HNO_3 (D) 9:6:3 HCl: H_2SO_4: HNO_3

58. Which law regulated the emission of air pollutants from auto mobiles?
(A) The Environment (Protection) Act 1986
(B) The Air (Prevention and control of pollution) Act,1981
(C) The Air (Prevention and control of Pollution) Rules,1982
(D) The Air (Prevention and Control of Pollution) Amendment Act, 1987

59. Under which law the standards for emission or discharge of environmental pollutants from specific industries were first set?
(A) The air (Prevention and Control of Pollution) Amend-ment Act,1987
(B) The Environment (Protection) Rules, 1986
(C) The Water (Prevention and Control of Pollution) Rules,1975
(D) The Environment (Protection) Act, 1986

60. Using which method is manure sample with more than 1% organic carbon determined?
(A) Chromic acid wet digestion method
(B) Dry combustion method
(C) Partial combustion method
(D) Walkley and Black method

61. What is the maximum permissible limit of BOD in industrial effluent that can be used for irrigation?
(A) 20 mg L^{-1} (B) 10 mg L^{-1}
(C) 30 mg L^{-1} (D) 100 mg L^{-1}

62. World Ozone day is celebrated on ________
(A) June 16 (B) July 16
(C) August 16 (D) Sept. 16

63. Who coined the term 'Smog'?
(A) Haeckal (B) Dr. Harold Des Veaux
(C) Odum (D) Tansley

64. The smell of H_2S is ________

(A) Rotten egg (B) Sweet smell

(C) Pungent (D) Obnoxious

65. **What is the maximum permissible limit of TDS in irrigation water?**

(A) 1000 mgl^{-1} (B) 2100 mgl^{-1}

(C) 30 mgl^{-1} (D) 6 mgl^{-1}

66. **Which industry that generates high amount of organic waste?**

(A) Distillery (B) Tannery

(C) Paper (D) Pharmaceutical

67. **The salvaging of usable metals, paper plastic and glass from municipal solid waste and sending them to manufacturing industries is called ________**

(A) Resource utilization (B) Resource recovery

(C) Resource conservation (D) Resource extension

68. **Which material can be recycled easily and creates minimum pollution?**

(A) Iron (B) Aluminium

(C) Glass (D) Plastics

69. **The first Act in India, passed in confirmation to the United Nations Conference on the human Environment Held in Stockholm in 1972 was ________**

(A) The Environment (Protection) Act 1986

(B) The water (Prevention and Control of Pollution) Act, 1981

(C) The Air (Prevention and control of pollution) Act, 1974

(D) The Water (Prevention and Control of Pollution) Act, 1981

70. **Which process produces high amount of green house gas?**

(A) Aerobic process (B) Anaerobic process

(C) Fermentation (D) None of the above

71. **What is the density of heavy metals?**

(A) 2g/cc (B) 5g/cc

(C) 2.5g/cc (D) 4g/cc

72. **Under the environment (Protection) Rules, 1986, bioassay test involving survival of fish is required for efflu-ents generated from which industry?**

(A) Oil refinery (B) Caustic soda

(C) Composite woollen mills (D) Thermal power plants

73. Which toxic material is released from the incineration of municipal solid wastes?
(A) HCL gas (B) Dioxins
(C) Heptachlor (D) Methyl mercury

74. Which country has the world's best waste management and recycling programme?
(A) USA (B) Switzerland
(C) Germany (D) Japan

75. Which of the following is the best way to handle food processing waste?
(A) Treat it in anaerobic digesters
(B) Treat it by aerobic treatment
(C) Burn it in incinerators
(D) Compost it in mixed form

76. The most heavily polluted river in India is ________
(A) Ganges (B) Cauvery
(C) Damodar (D) Narmada

77. A report on "Global Environmental Monitoring System" in 1975 was published by ________
(A) EPA (B) WHO
(C) FAO (D) UNEP

78. The use of DDT in India for agricul-tural use is banned in the year ________
(A) 1980 (B) 1985
(C) 1986 (D) 1984

79. Cadmium poisoning leads to ________
(A) Itai Itai disease
(B) Knock knees syndrome
(C) Methemoglobinemia
(D) Cancer

80. The highest cess charges under Water Prevention and control of pollution cess Act, 1978 is for which category of purpose?
(A) Domestic use of water being supplied by municipality
(B) Water getting contaminated with biodegradable pollutants
(C) Water getting contaminated with non-biodegradable and toxic pollutants
(D) Industrial cooling spraying in mine piles or boiler feed

81. The Centre for Environmental Education (CEE) is located at __

(A) New Delhi (B) Ahmedabad
(C) Coimbatore (D) Bhopal

82. Which of the following is not related to United Nations?

(A) WHO (B) SACEP
(C) UNEP (D) Earthscan

83. What is the purpose of the ENVIS centers opened by Ministry of Environment?

(A) Disseminate environmental information
(B) Carry out environmental research
(C) Impart training to students
(D) Spread awareness among masses about environment

84. Which is the most suitable media for environmental education in remote villages?

(A) Television (B) Newspaper (Local)
(C) Radio (D) Handbills

85. Lead inhalation leads to ____________

(A) Anemia (B) Berylliosis
(C) Black lung (D) White lung

86. The main source for cadmium pollution in soil is _______

(A) Nitrogenous fertilizers
(B) Phosphotic fertilizers
(C) FYM application
(D) Potassium fertilizers

87. The efficiency of bioremediation technique mostly depends on _______

(A) Total quantity of heavy metals present
(B) Bio availability of metals
(C) Type of microorganism
(D) Type of soil

88. Heavy metals inhibit the organisms by affecting ___________

(A) Growth (B) Enzyme synthesis
(C) Protein synthesis (D) Amino acid synthesis

89. Which equipment releases Ozone gas?

(A) Gas stove (B) Refrigerator
(C) Photocopier (D) Microwave oven

90. **Which person is qualified for appointment as a Government Analyst under Environment (Protection) Rules 1986?**
(A) A post-graduate in environmental science with one years experience
(B) A Graduate in environmental testing or analysis
(C) A graduate in science with four years of experience in environmental testing or analysis
(D) A post-graduate in science with one year experience in environ-mental testing or analysis

91. **The toxic form of mercury is ________**
(A) Oxidized form
(B) Mercurous acid
(C) Elemental mercury
(D) Methyl mercury

92. **The disease caused by cotton fiber in the air is ________**
(A) Gout
(B) Chronic bronchitis
(C) Byssinosis
(D) Black lung cancer

93. **Which of the following is a chemical fertilizer that washes out into the water supply in large amounts?**
(A) Nitrite
(B) Nitrate
(C) Sulphates
(D) Phosphates

94. **The more toxic gaseous components released due to burning of fossil fuel is ________**
(A) CO (B) CO_2 (C) SO_2 (D) NO_2

95. **With the increase in temperature of the receiving water, the ________**
(A) DO content decreases
(B) DO content increases
(C) Demand of Oxygen increases
(D) Demand of Oxygen increases

96. **Which radioactive element is an Indoor pollutant?**
(A) O-16
(B) N-13
(C) Carbon -14
(D) Radon

97. **The major pollutant present in cigarette smoke is ________**
(A) CO and Dioxin
(B) Nicotine and CO
(C) Benzene and CO
(D) Dioxin and Benzene

98. **An alternative of chrome tanning of leather is ________**
(A) Sulphate tanning
(B) Vegetable tanning
(C) Chlorine tanning
(D) Sulphite tanning

99. Which mode of transportation is least damaging to the environment?
(A) Buses (B) Electric trains
(C) Aeroplane (D) Cars

100. Which country releases maximum of the green gases?
(A) India (B) China
(C) United States (D) Former Soviet Union

101. 1 Du (Dobson Unit) at standard temp Pressure is equal to Ozone column thickness of ______
(A) 0.1 mm (B) 1.0 mm
(C) 0.01 mm (D) 0.001 mm

102. The gas responsible for green houses effect is ______
(A) Ozone (B) CO_2
(C) CO (D) H_2S

103. Lead poisoning affects the function of
(A) Liver (B) Kidney
(C) Lungs (D) Heart

104. Ozone at ground surface is responsible for the formation of a hazardous pollutant namely ______
(A) Smoke (B) Fog
(C) Smog (D) None of the above

105. The major element responsible for photochemical smog is ______
(A) SO_2 (B) CO
(C) CO_2 (D) NO_2

106. Which part of the human body is the first to be affected by nuclear radiation?
(A) Lungs (B) Brain
(C) Bone Marrow (D) Liver

107. In the absence of gases like CO_2, NO_2 and water vapour which keeps the earth's surface warm, the average surface temperature of earth would be ______
(A) 18°C (B) 8°C
(C) -18°C (D) 0°C

108. The biggest source of atmospheric hydrocarbon is ______
(A) Refuse burning (B) Petroleum burning
(C) Coal burning (D) Power generation

109. When methanol is used as a fuel for automobile emission, which pollutant increases significantly?
(A) NOx (B) CO
(C) Formaldehyde (D) PAN

110. Three Mile Island nuclear power plant is located in ________
(A) Former USSR (B) USA
(C) France (D) Germany

111. The most abundant fossil fuel is ________
(A) Coal (B) Oil
(C) Natural gas (D) Peat

112. In tricking filters, the organism that grows in large population is ________
(A) Zooglea (B) Algae
(C) Planktons (D) None of the above

113. Laccase is used to degrade ________
(A) Sugar Compounds (B) Sulphides
(C) Lignin (D) Phenol compounds

114. The unit of odour is ________
(A) mg/kg of air (B) g/m^3
(C) E/m^3 (D) No unit

115. Which of the following has the maximum ozone depleting potential?
(A) Halon 1301 (B) HCFC22
(C) CFC 115 (D) CFC 12

116. Which chemical species has the least Ozone depleting potential?
(A) Halon 2402 (B) HCFC 22
(C) CFC 11 (D) CFC 113

117. Amendments to the Montreal protocol were undertaken in 1990 at ________
(A) Paris (B) London
(C) Tokyo (D) Amsterdam

118. IPCC is concerned with ________
(A) Ban on ODS
(B) Climate change
(C) Biodiversity conservation
(D) Trade of endangered species

119. The National Forest Policy was first adopted in 1894, revised in 1952 and was again revised in the year ______

(A) 1980 (B) 1974
(C) 1988 (D) 1972

120. Sago industry waste water is rich in ______

(A) Cellulose (B) Phenol
(C) Salt (D) Starch

121. The biodegradation ability of pseudomonas is attributed to ______

(A) The presence of plasmids
(B) Resistance to adverse conditions
(C) Presence of hydroxylase enzyme
(D) None of the above

122. Which one of the following is not a suspended cell aerobic treatment process?

(A) Activated sludge process
(B) Oxidation ponds
(C) Trickling filters
(D) Aerated lagoon

123. The color of the paper mill effluent waste water is due to ______

(A) Phenolic compounds
(B) Cholrides
(C) Sulphates
(D) Lignocellulosic compounds

124. What is the maximum permissible E coli count in drinking water?

(A) 10/ml (B) 2/ml
(C) 5/ml (D) 7/ml

125. The color of distillery effluent is due to ______

(A) Lignocellulosic compounds
(B) Caramel
(C) Iron oxides
(D) Dextrose

126. The microorganism used for the remediation of oil spills is ______

(A) Pseudomonas putida (B) Alcaligenes eutrophus
(C) Bacillus (D) Phanerochaete

127. The major problem in paper and pulp industry effluent is ________

(A) BOD (B) Phenols
(C) Chlorides (D) Cellulose

128. Which of the following is used to assess the degree of water pollution due to sewage water?

(A) *E. coli* count (B) COD
(C) BOD (D) Salt content

129. The tolerance limit of endosulfan in fruits and vegetables is ________

(A) 0.5 ppm (B) 1.0 ppm
(C) 1.5 ppm (D) 2.0 ppm

130. The odour in the paper and pulp waste industry is due to ________

(A) Dioxins (B) Mercaptans
(C) Sulphides (D) Ammonia

131. Which microorganism is used in biodegradable plastic production?

(A) *Pseudomonas putida* (B) Alcaligenes eutrophus
(C) Spirulina (D) Zoogles

132. Electrostatic precipitators are used to remove ________

(A) Heavy metal (B) Sulphur compounds
(C) Particulates (D) Charges particles

133. Biological treatment of wastewater aims to remove ________

(A) Inorganic pollutants
(B) Organics and suspended solids
(C) Heavy metals
(D) None of the above

134. The principle in estimation of dissolved oxygen is ________

(A) Iodometric (B) Gravimetric
(C) Colorimetric (D) Potentiometric

135. Hospital waste is disposed by________

(A) Composting (B) Landfill
(C) Fermentation (D) Incineration

136. The commonly used method for sewage treatment in India is ________

(A) Oxidation ponds
(B) Trickling filters
(C) Activated sludge process
(D) Rotating biological contactor

137. Black lung disease occurs among ______
(A) Quarry workers (B) Refinery workers
(C) Coal miners (D) Smokers

138. Van Allen belt is a zone of ______
(A) CO_2 (B) O_2
(C) O_3 (D) CO

139. The final product of the continuous disintegration of Uranium 238 is ______
(A) Copper 206 (B) Lead 206
(C) Chromium (D) Zinc

140. Which among the following is an isotope of hydrogen?
(A) Uranium (B) Thorium
(C) Deuterium (D) Hydronium

141. Brown smoke refers to smoke produced by burning of ______
(A) Coal (B) Wood
(C) Garbage (D) Tyres

142. The loss of which of the following inhibitor in MIC tank was responsible for Bhopal gas tragedy?
(A) HCl (B) H_2SO_4
(C) HNO_3 (D) Phosgene

143. Silo-fillers disease in man is due to ____
(A) N_2O poisoning (B) NO_2 Poisoning
(C) CO Poisoning (D) CO_2 Poisoning

144. Which process/activity doesn't release any greenhouse gas?
(A) Livestock rearing
(B) Paddy cultivation
(C) Clear cutting of forests
(D) Nuclear energy generation

145. The Chemical formula of CFC-11 and CFC-12 is ______
(A) CFCL3 and CF2CL2
(B) CF3Cl and CF2 CL2
(C) CF2CL2 and CF3 Cl
(D) CF2Cl2 and CF Cl3

146. The largest use of CFCs in 1985 was as ________
(A) Refrigerants (B) Rigid foam insulation
(C) Aerosol propellants (D) Solvents

147. The use of hydro fluorocarbons in place of chlorofluorocarbons is presently advocated as they ________
(A) Don't reach to stratosphere after use
(B) Don't deplete Ozone layer
(C) Have very low Ozone layer depleting potential
(D) Don't degrade to release chlorine atoms in stratosphere

148. Which chemical has the maximum life in natural environment?
(A) CFC 12 (B) CFC 115
(C) HCFC 22 (D) Halon 1211

149. Pyrheliometer is used in measuring ________
(A) Sun spots (B) Solar radiation
(C) Air temperature (D) Temperature of plants

150. Objectives and guiding principles of environmental education were first formulated in 1977 at which place?
(A) Stockholm (B) Tbilisi
(C) Rio De Janeiro (D) New Delhi

151. General circulation models (GCMs) are used to ________
(A) Predict climate
(B) Predict cyclones
(C) Study ocean currents
(D) Study effects of temperature on wind

152. The book "Silent Spring" published in 1960 was written by ________
(A) Sunderlal Bahuguna (B) Rachel Carson
(C) Salim Ali (D) David Matthews

153. Joint Forest management was initiated in India in 1972 at ________
(A) Bastar, Chhattisgarh
(B) Arabari, West Bengal
(C) Pithoragarh, Uttaranchal
(D) Connor, Tamil Nadu

154. The compound that causes egg shell thinning in birds is ________
(A) DDT (B) Mercury
(C) BHC (D) PCB

155. The compound that causes chromosomal malformation is ________

(A) Dioxin (B) Benzene
(C) Aldrin (D) None of the above

156. Eutrophication is caused due to ______

(A) Nitrates (B) Phosphates
(C) Both A and B (D) Neither A or B

157. According to Irving Williams order, the heavy metal that is adsorbed first in the mineral surface is _______

(A) Lead (B) Mercury
(C) Copper (D) Zinc

158. In anaerobic condition the heavy metals are precipitated as ________

(A) Carbonates (B) Phosphates
(C) Hydroxides (D) Sulphides

159. During heavy metal complexa-tion the electron donor is _____

(A) Oxygen (B) Nitrogen
(C) Sulphur (D) All the above

160. Hexavalent chromium is toxic because of its _______

(A) Oxidizing potential (B) Mobility
(C) Solubility (D) Reducing potential

161. Keshan's disease in humans is caused due to _______

(A) Chromium (B) Nickel
(C) Selenium (D) Lead

162. Biologically non-essential heavy metal is ________

(A) Copper (B) Cadmium
(C) Zinc (D) Cobalt

163. The maximum permissible level of selenium in soil is ________

(A) 0.5mg/kg (B) 1.0mg/kg
(C) 0.1mg/kg (D) 0.05mg/kg

164. The accumulation level of hyperaccumulators is ________

(A) 50-75 mg/kg (B) 75-100 mg/kg
(C) 50-100 mg/kg (D) 100-1000 mg/kg

165. The introduction of specific microbe in situ for bioremediation of contami-nated sites is ____________

(A) Biostimulation (B) Bioaugmentation
(C) Bioaccumulation (D) Biomagnification

166. Emphysema is caused due to ______

(A) CO_2 (B) CO
(C) NO_2 (D) SO_4

167. The principle in using activated Carbon for wastewater treatment is ______

(A) Coagulation (B) Absorption
(C) Adsorption (D) All the above

168. The level of CO_2 in atmosphere is ______

(A) 0.30% (B) 0.03%
(C) 0.003% (D) 0.1%

169. The largest bioremediation project under taken so far to treat oil spill was at ______

(A) Price Island, Philippines
(B) Prince William sound, Alaska
(C) Kuwait shoreline
(D) Exxon Valdez, Alaska

170. Polyhydroxybutyrate (PHB) can be extracted from ______

(A) Alcaligenes eutrophus
(B) Aspergillus niger
(C) Penicillum spp
(D) Phanerochaete chrysogenum

171. Which one among the following is a major indoor air pollutant in India?

(A) Ozone (B) Radon
(C) CO (D) Formaldehyde

172. Implementing ambient air quality standards in a region can be classified into how many categories?

(A) Three (B) Four
(C) Five (D) Six

173. Sodium tetra chloro mercurate method mentioned in Indian standards is used for estimating which air pollutant?

(A) CO (B) SO_2
(C) NO_2 (D) NOx

174. Rotating biological contactors are used for ______

(A) Reducing BOD of wastewater
(B) Biofiltration of water
(C) Removing colour of textile effluents
(D) Removing pathogens from sewage

175. Fine bubble air diffusers are used in ________
(A) Trickling filter
(B) Activated sludge process
(C) Rotating biological contractor
(D) Oxidation pond

176. ________ is an air pollution control device that removes particulates out of air or gas released from commercial processes or combustion for electricity generation
(A) Cyclone seperator (B) Venturi scrubber
(C) Bag filter (D) Electrostatic precipitator

177. Which among the following device has dust collection efficiency greater than 99%?
(A) Wet electrostatic precipitator
(B) Venture scrubber
(C) Cyclonic spray scrubber
(D) Multicyclone

178. Which compound is in highest concentration among the following in cigarette smoke?
(A) SO_2 (B) NOx
(C) Furans (D) Pyridine

179. Minamata disease is caused due to ________
(A) Phenyl mercury (B) Inorganic mercury
(C) Methyl mercury (D) Benzyl mercury

180. In urban areas, Ozone is found in trophosperic zone due to ________
(A) Vehicular traffic (B) Deforestation
(C) Global warming (D) Change in surface Albedo

181. Super bug is ________
(A) *Bacillus thuringiensis* (B) *Penicillium notatum*
(C) Pseudomonas putida (D) *Streptomyces sp.*

182. Which type of waste is contained in "Love canal" dump site?
(A) Hospital (B) Municipal waste
(C) Nuclear waste (D) Chemical waste

183. Which is not considered a sensitive area for implemen-ting ambient air quality standards?
(A) Hospital (B) Tourist resorts
(C) National Monuments (D) National Parks

184. Oxygen and Ozone combined can absorb solar radiation of wavelength ______
(A) Between 3-6 m (B) Less than 3 m
(C) Less than 1 m (D) Less than 0.3 m

185. The most obvious toxic product of combustion of plastic is ____
(A) SO_2 (B) CO
(C) CO_2 (D) Dioxins

186. Which among the following is a major atmospheric sink for CH_4 and CO?
(A) Hydroxyl ions (B) Oxides
(C) Water vapor (D) Hydrogen ions

187. Mesothelioma is a type of cancer which occurs due to ______
(A) Asbestos (B) Copper
(C) Silver (D) Iron

188. Methemoglobinemia is caused due to ______
(A) Nitrates (B) Phosphate
(C) Lead (D) Iron

189. The distinct smell of sea is due to ______
(A) Hydrogen sulphide (B) Ammonia
(C) Methane (D) Dimethyl sulphide

190. The removal of odour from polluted air is by ______
(A) Biofilters (B) Cyclotron
(C) Electrostatic Precipitators (D) Bioscrubbers

191. Carbogen is a mixture of ______
(A) 95% O_2 & 5% CO_2 (B) 95% CO_2 & 5% O_2
(C) 50% CO_2 & 50% O_2 (D) 75% CO_2 & 25% O_2

192. Heavy duty diesel powered vehicles contribute more of ______
(A) NO_2 and Particulate matter
(B) CO & HC
(C) CO & CO_2
(D) SO_2 & NO_2

193. What fraction of the mass of atmosphere does troposphere and stratosphere together account for?
(A) 89.3% (B) 76.3%
(C) 83.9% (D) 99.9%

194. **The boundary between troposphere and stratosphere is called ______**

(A) Ionosphere (B) Mesosphere
(C) Thermosphere (D) None of the above

195. **Among the following organisms, the first rate pollution monitor is ______**

(A) Honey bees (B) Animals
(C) Butterflies (D) Molluscs

196. **Green house effect was discovered by ______**

(A) Joseph Fourier (B) de Voeux
(C) Charles Dufay (D) Vladimir M. Komarov

197. **A report on "Global Environmental Monitoring System" in 1975 was published by ______**

(A) EPA (B) WHO
(C) FAO (D) UNEP

198. **DDT has been banned because of its ______**

(A) Very high toxicity (B) Wide spectrum activity
(C) Costly (D) Residual effect

199. **Eye irritation while traveling by vehicle is due to ______**

(A) Diesel (B) Lead bromide
(C) Particulate matter (D) None of the above

200. **The central and state Boards for prevention and control of pollution were first constituted under which law?**

(A) The Environment (Protection) Act, 1986
(B) The Air (Prevention and Control of Pollution) Act, 1981
(C) The Water (Prevention and Control of Pollution) Act, 1974
(D) The water (Prevention and Control of Pollution) Cess Act, 1978

201. **When a person or local authority liable to pay cess under the water (Prevention and Control of Pollution) Cess Act 1978, installs any plant for treatment of sewage or trade effluents then he or the concerned local authority is entitled for a rebate of ______on cess payable**

(A) 50% (B) 70%
(C) 33% (D) 60%

202. **The linear roll back used for air quality modeling is a ______**

(A) Deterministic model (B) Receptor - oriented model
(C) Empirical statistical mode (D) Time series model

203. Mitti Bachao Abhiyan movement aims at ________
(A) Save soil (B) Deforestation
(C) Air protection (D) Water protection

204. In anaerobic chromium reduction, the electron acceptor is ________
(A) Sulphate (B) Organic matter
(C) H_2S (D) CO_2

205. The silent killer gas that combines with hemoglobin in the blood is ________
(A) CO_2 (B) SO_2
(C) H_2S (D) CO

206. Lichens have disappeared from cities because they are highly susceptible to ________
(A) Carbon dioxide (B) Lead
(C) Nitric Oxide (D) Sulphur dioxide

207. The maximum greenhouse contribution in earth's temperature is by ________
(A) CH_4 (B) CO_2
(C) N_2O (D) CFCs

208. Which of the following is true when CO_2 and CH_4 are compared in terms of greenhouse efficiency?
(A) CH_4 is 16 times more efficient than CO_2
(B) CO_2 is 26 times more efficient than CH_4
(C) CH_4 is twice efficient than CO_2
(D) CH_4 is 25 times more efficient than CO_2

209. Mauna Loa, in Hawaii is famous for ________
(A) Botanical garden
(B) Monitoring sea level rise since 1950
(C) Biggest collection of mammals fossils
(D) Continuous monitoring of atmo-spheric CO_2 since 1957

210. Notification on environmental Impact Assessment of Development Projects is covered under ________
(A) Environmental Impact assessment Act
(B) Environment (Protection) Act
(C) The Factories Act
(D) The Environment Conservation Act

211. Which part of human body is the first to be affected due to vehicular pollution?

(A) Kidney (B) Liver
(C) Lungs (D) Brain

212. The analytic technique suitable for analysis of aromatic hydrocarbon is ______

(A) Spectrophotometry
(B) Nephelometry
(C) Gas chromatography
(D) Atomic absorption spectroscopy

213. The air pollutant "Acrolein" is of type ______

(A) Aldehyde (B) Ketone
(C) Paraffin (D) Olefin

214. Which tissue is the best indicator of lead accumulation in human body?

(A) Fatty tissues (B) Bones
(C) Brain (D) Blood

215. Which compound is used for making mineral water bottles?

(A) Polyvinyl chloride
(B) High density polyethylene
(C) Low density polyethylene
(D) Polyethylene terephthalate

216. An example of secondary pollutant is ______

(A) CFCS's (B) CH_4
(C) PAN (D) CO

217. A pure or nearly pure water contains a BOD of approximately (mg/Lit) ______

(A) 30 (B) 20-30
(C) 0-3 (D) 10-12

218. Which of the following conversion is not possible in the catalytic converter used in motor vehicle?

(A) NOX to N_2 (B) Hydrocarbons to CO_2
(C) CO to CO_2 (D) SO_2 to Sulphur

219. M 85 is a ______

(A) Plant oil similar to diesel discovered in 1985
(B) A mixture of 85% methanol and 15% petrol
(C) The most fuel efficient car developed so far
(D) Japanese technology capable of reducing NOx emissions by 85%

220. Which type of scrubber is most suitable for controlling fine particulate emissions?
(A) Mechanical scrubber (B) Venturi scrubber
(C) Spray tower (D) Packed tower

221. The photochemical smog formation in urban area is mainly due to the presence of __________ in the atmosphere
(A) Ozone, PAN and Nitrogen Dioxide
(B) Fog and Smog
(C) Particulates and fog
(D) Winter climate and high humidity

222. Which one of the following additives is mixed in petrol to increase its oxygen content?
(A) Methyl tertiary butyl ether (MTBE)
(B) Tetraethyl lead
(C) Propyl alcohol
(D) Methyl ethyl lead

223. Thousands of people died due to toxicity of __________ in Bhopal gas tragedy in 1984.
(A) Pesticide fumes and CO
(B) Methyl isocyanate carbon monoxide
(C) Methyl isocyanate and phosgene
(D) Methyl isocyanate and alkaline

224. While establishing an air pollution generating industry, the maximum mixing depth of the area should be __________ to minimize air pollutants buildup
(A) 1000m (B) >2000m
(C) >5000m (D) <5000m

225. The solubility of gases in water can be calculated using________
(A) Henry's law (B) Raoult's law
(C) Avogadro's law (D) Gay lussac's law

226. Which organ of human body does Stontium-90 affect?
(A) Muscles (B) Brain
(C) Bones (D) Kidneys

227. The two formulas recommended by Central pollution Control Board for deciding the minimum stack height are based on emission rates of________
(A) CO_2 and soot (B) SO_2 and CO_2
(C) SO_2 and NO_2 (D) SO_2 and particulate matter

228. Which one among the following is not a method of air pollution control at source?
(A) Raw material changes (B) Scrubbing of flue gases
(C) Change in process (D) Modification in equipments

229. The cleanest fuel among the following is ______
(A) Coke (B) Gasoline/Petrol
(C) Natural Gas (D) Diesel

230. Which harmful pollutant is formed when elements react with oxygen and water in mining deeper deposits?
(A) Sulphates (B) Oxides
(C) Hydroxides (D) Sulphides

231. Which agency has the power of setting pollution standards and enforcing them?
(A) UNEP (B) WHO
(C) EPA (D) CPCB

232. Which scientist detected acid in rainwater for the first time?
(A) Ravish Malhotra (B) Yuri Gagarin
(C) Henry Becquerel (D) Robert Angus smith

233. Atomic Absorption Spectroscopy is used to analyze ______
(A) Heavy elements
(B) Dissolved organic compounds
(C) Dissolved gases
(D) Particle size

234. In automobile exhaust, lead is emitted mostly as ______
(A) Tetra alkyl lead (B) Metallic lead
(C) Lead sulphide (D) Lead bromochloride

235. Which among the following is the biggest source of air pollution?
(A) Thermal power plant (B) Industries
(C) Transportation (D) Refuse disposal

236. The air flow-rate through the high volume air sampler is normally maintained at ______ during ambient air quality monitoring.
(A) 1.7 m^3/min (B) 6.3 m^3/min
(C) 17 m^3/min (D) 20.2 m^3/min

237. The pH of acid rain will be ______
(A) 7.0 (B) 8.5
(C) 5.6 (D) 6.7

238. **Which toxic heavy metal is present in the effluent from modern tanneries?**

(A) Zn (B) Ni
(C) Cr (D) Pb

239. **An example of a PAH compound is ______________**

(A) Peroxyacetyl nitrate (PAN)
(B) Tetra ethyl lead (TEL)
(C) Benzene hexachloride (BHC)
(D) Benzo (or) pyrene

240. **Carbon monoxide gas affects humans by forming which compound in the blood?**

(A) Bicarbonate (B) COHb
(C) $HbCO_3$ (D) H+

241. **Which instrumental method is used to estimate NO_2?**

(A) Flame photometer
(B) Atomic absorption spectroscopy
(C) Chemiluminescence
(D) High performance liquid chromatography

242. **The most commonly used coagulant in wastewater treatment is ________**

(A) Chlorine (B) Acids
(C) Alum (D) Carbon do oxide

243. **Hardness of water is due to the presence of ________**

(A) Calcium and Magnesium (B) Sodium
(C) Potassium (D) Phosphorus

244. **The bioindicator organism for the fecal contamination of water is ____________**

(A) *Bacillus* sp. (B) *Clostridium* sp.
(C) Algae (D) *E. Coli*

245. **The transfer of contaminants through food chain is referred as ________**

(A) Biotransformation (B) Biomagnification
(C) Biological clocks (D) Biomonitoring

246. **Which one of the following is not a major source of soil pollution?**

(A) Coal mining (B) Coal based Thermal power plants
(C) Land filling of municipal waste (D) Textile units

247. Based on ____________ of pollutants, pollution is classified as pesticide pollution, plastic pollution, heavy metal pollution, radiation pollution, oil pollution, sewage pollution, noise pollution, etc.

(A) Nature (B) Sources
(C) Environment (D) Concentration

248. Factors that can shape climate are called ________

(A) Climate forcing (B) Climate change
(C) Climate drive (D) Weather drive

249. Climate system that responds more slowly to climate forcing are

(A) Ocean and ice caps (B) Mountain
(C) Buildings (D) All

250. Example for internal climate forcing mechanism

(A) Thermohaline circulations
(B) Changes in solar output
(C) Changes in CO_2 concentration
(D) Changes in water vapour

251. The response of the climate system will be fast, if

(A) a sudden cooling due to airborne volcanic ash reflecting sunlight
(B) thermal expansion of warming ocean water
(C) sudden loss of albedo in the arctic ocean as sea ice melts
(D) thermal expansion of the water

252. The fundamental part of climate system

(A) Soil (B) Ocean
(C) Mountains (D) Volcanoes

253. The eruption of Mount Pinatubo in 1991, the second largest terrestrial eruption of the 20th century, affected the climate substantially, subseque-ntly global temperatures decreased by about

(A) 0.1 °C (B) 0.5 °C
(C) 1.0 °C (D) 5.0 °C

254. the study of contemporary and fossil palynomorphs, including pollen

(A) Polynology (B) Pollenology
(C) Palynology (D) Seismology

255. Dendroclimatology is the analysis of ________ to determine past climate variations
(A) Tree pollen
(B) Fossil palynomorph
(C) tree ring growth patterns
(D) ice core

256. The Intergovernmental Panel on Climate Change (IPCC) was established by
(A) World Meteorological Organiza-tion (WMO)
(B) World Health Organization (WHO)
(C) Environmental Protection Agency (EPA)
(D) Earth watch

257. Which among the following is a clean technology to reduce emission?
(A) Carbon Fuel (B) Hydrogen fuel
(C) Sulphur fuel (D) CNG

258. FACE denotes
(A) Free Air Chemical Enrichment
(B) Free Air CO_2 Enrichment
(C) Free Air Carbon Enhancement
(D) Free Air Carbon Energy

259. Most of the warming observed over the past 50 years is attributable to
(A) Human activities (B) Volcanic eruption
(C) Forest fires (D) Continental drift

260. The anticipated change in global temperature of ________ or more degrees could be devastating, leading to significant maladaptations over the next rotation cycle of most tree species.
(A) 3.0 to 5.0 °C (B) 0.5 to 1.0 °C
(C) 0 to 0.5 °C (D) 1.0 to 1.5 °C

261. Which among the following is not a component of sustainable forest framework
(A) Environment (B) Economic
(C) Social (D) Political

262. Which among the following is not considered as carbon sink
(A) Saline aquifer (B) Ocean water
(C) Aging oil fields (D) Atmosphere

263. Trees play a key role in maintaining soil fertility because
(A) Their leaves rot down and fertilize the soil
(B) They capture the atmospheric Nitrogen
(C) They live for 100 years
(D) They absorbs more water

264. _______can reduce the damaging effects of rising temperatures associa-ted with climate change naturally
(A) Glaciers (B) Ocean
(C) Trees (D) Soil

265. _________ is the capacity for a socio-ecological system to absorb stresses and maintain function in the face of external stresses imposed upon it by climate change
(A) Clima adopt
(B) Climate resilience
(C) Climate vulnerability
(D) Climate forcing

266. Global warming is mainly because of the accumulation of _________
(A) Short wave lengthy radiation
(B) Long wave length radiation
(C) Medium wave length radiation
(D) Both short and long wave length radiation

267. Climate resilience efforts encompass _________
(A) Social (B) Economical
(C) Political (D) All

268. The following famous quote "Environmental Pollution is an incurable disease. It can only be prevented." is written by _________
(A) Mahatma Gandhi (B) Barry Commoner
(C) A.G. Tanlsey (D) Odum

269. Earth has enough for everyone's need, but not enough for everyone's greed" The famous quote said by _______
(A) Dr. B.R. Ambedkar (B) Jawaharlal Nehru
(C) Narendra Modi (D) Mahatma Gandhi

270. Swatch Bharat Mission in India was launched by _________
(A) Narendra Modi (B) Jawaharlal Nehru
(C) V. P. Singh (D) Manmohan Singh

271. ___________ is a substance which does not occur naturally; introduced as a result of anthropogenic activities.
(A) Recalcitrant (B) Pollutant
(C) Contaminant (D) All the three

272. Temperature shoots up in water body due to the discharge of hot water is known as ________
(A) Acid rain (B) Light pollution
(C) Thermal pollution (D) Noise pollution

273. Modern Agricultural practices does not involves ________
(A) Application of chemical fertilizers
(B) Application of pesticides
(C) Application of Herbicides
(D) Application of Vermicompost

274. Enrichment of water bodies due to the addition of nutrients are called ________
(A) Biomagnification (B) Bioaccumulation
(C) Eutrophication (D) Hypoxia

275. Which among the following micro-organisms is considered as indicator of water pollution and faecal contamination?
(A) *Bacillus* (B) *E. coli*
(C) *Escherichia coli* (D) Both B and C

276. Which among the following sector contributes highest share of air pollution in India?
(A) Industries (B) Transportation
(C) Educational Institutions (D) Hospitals

277. Land fill sites generally produce ______
(A) Methane and CO_2 (B) CO and CO_2
(C) SO_2 and NO_2 (D) CFC and Methane

278. Uncontrolled burning of water is known as ______
(A) Open burning (B) Incineration
(C) Pyrolysis (D) Gasification

279. Stoves and cookers produces ______
(A) CO and NO_2 (B) SO_2
(C) SO_4 (D) CFC

280. The recent Delhi smog 2018 was due to ______
(A) Burning of paddy straw (B) Burning of maize straw
(C) Burning of wheat stubbles (D) All

281. Live stock and poultry industry wastes produces ________

(A) CFC and Ozone (B) Ozone and Methane
(C) CH_4 and NH_3 (D) SO_2 and NO_2

282. Which among the following are examples of geogenic contaminant in India?

(A) Fe, P, As, F (B) Fe, Cl, Mn, Zn
(C) Cr, Ni, Cd, Hg (D) Cu, Zn, Fe, Mn

283. Leaching of toxic inorganic material from the underlying parent rock materials are known as

(A) Thermal pollution (B) Geogenic pollution
(C) Nuclear pollution (D) Radioactive pollution

284. Application of phosphotic fertilizers leads to the accumulation of ________

(A) Mercury (B) Cadmium
(C) Chromium (D) Lead

285. Vellore of Tamil Nadu, India is known for its ________

(A) Dye industries (B) Textile industries
(C) Tannery industries (D) Sago industries

286. Tiruppur of Tamil Nadu, India is known for ________

(A) Dye industries (B) Sago industries
(C) Tannery industries (D) Foundries

287. Salem, Tamil Nadu is known for ________

(A) Sago industries (B) Tannery industries
(C) Dye industries (D) Textile industries

288. The issue of Uranium poisons in Punjab in the year 2009 was due to ________

(A) Fly ash of thermal power stations
(B) Nuclear test (C) Oil refineries
(D) Collision of submarine is the Arabian coast.

289. The WHO prescribed limit of PM 2.5 is ________

(A) 25 g/m^3 (B) 50 g/m^3
(C) 75 g/m^3 (D) 100 g/m^3

290. EPI denotes

(A) Environmental index
(B) Environmental performance index
(C) Environmental purity index
(D) Environmental population index

291. Which among the following city is considered as gas chamber in India due to air pollution?

(A) Goa (B) Delhi
(C) Mumbai (D) Hyderabad

292. Every drop in the ocean counts, quote by

(A) Yoko onto (B) Yoko Anto
(C) Tansley (D) Odum

293. "War for ______: Get ready for the battle", famous quote caution us

(A) Water (B) Air
(C) Soil (D) All

294. Compound used as water proofing agents

(A) Cr (VI) (B) As (III)
(C) As (V) (D) PCB

295. Presence of __________ in water causes typhoid

(A) *Salmonella typhi* (B) *Salmonella paratyphi*
(C) *Shigella* sp (D) *E. coli*

296. Black foot disease is caused by ________

(A) As (B) Cd
(C) Ni (D) Hg

297. Itai Itai / Ouch Ouch disease is caused by _________

(A) As (B) Cd
(C) Ni (D) Hg

298. Which among the following form of chromium is toxic to human beings?

(A) Cr (III) (B) Cr (VI)
(C) Cr (III) and Cr (VI) (D) Cr (II)

299. Presence of fluoride prevent tooth decay when the concentration is below ______ in water

(A) 5 mg L^{-1} (B) 4 mg L^{-1}
(C) 3 mg L^{-1} (D) 1 mg L^{-1}

300. Fluoride causes bone damage at _________

(A) 2 mg L^{-1} (B) 3 mg L^{-1}
(C) 4 mg L^{-1} (D) 5 mg L^{-1}

301. Knock knee syndrome in human was due to

(A) Chloride (B) Fluoride
(C) Hg (D) As

302. Minamatta disease caused by ______
(A) Hg (B) Ni
(C) Cd (D) Fe

303. In 'Eutrophication' 'Eu' means
(A) Less (B) More
(C) Small (D) None

304. The word Eutrophication is derived from the ________ word eutrophos means "Well nourished"
(A) Greek (B) Latin
(C) English (D) French

305. Terror of Bengal
(A) Water hyacinth (B) Typha
(C) Phragmetis (D) Lilly

306. Eutrophication in water is mainly caused by ______
(A) Chloride and sulphate (B) Ca and Mg
(C) PO_4 and NO_3 (D) CO and CO_2

307. Which among the following statements are correct?
(A) Phosphorous is generally less soluble than nitrogen
(B) Phosphorus leaching in soil is much slower than nitrogen
(C) Both A and B
(D) 'A' alone

308. The most active stage of lake usually occur during
(A) Rainy season (B) Winter
(C) Summer (D) Autumn

309. The nutrient rich top layer of lake is known as ______
(A) Epilimnion (B) Hypolimnion
(C) Metalimnion (D) Thermocline

310. The bottom layer of lake
(A) Epilimnion (B) Hypolimnion
(C) Metalimnion (D) Thermocline

311. The middle layer of lake
(A) Epilimnion (B) Hypolimnion
(C) Metalimnion (D) Thermocline

312. Warmest layer of lake
(A) Epilimnion (B) Hypolimnion
(C) Metalimnion (D) Thermocline

313. Some of the lakes have the tendency to become less nutrient rich with time. This reverse process is known as
(A) Eutrophication (B) Mesotriphication
(C) Meiotrophication (D) All the three

314. Total phosphorous concentration of most unpolluted fresh water will be in the range of ________
(A) 200 to 500 mg L^{-1} (B) 100 to 1000 mg L^{-1}
(C) 10 to 50 mg L^{-1} (D) < 10 mg L^{-1}

315. Red field ratio is connected with
(A) C:P is water body (B) N:P is water body
(C) O:P is water body (D) C:O is water body

316. Riparian buffer zones are
(A) Phosphorous is generally less soluble than nitrogen
(B) Phosphorus leaching in soil is much slower than nitrogen
(C) Both A and B
(D) 'A' alone

317. Altering the food web to restore the lake ecosystems health is known as ________
(A) Biomagnification (B) Biomanipulation
(C) Bioremediation (D) Bioaccumulation

318. Nitrate contamination is drinking water causes
(A) Blue baby syndrome (B) Methamoglobinemia
(C) Itai Itai (D) Both A and B

319. Mottling teeth is due to the presence of ________ is water
(A) Arsenic (B) Fluorides
(C) Chlorides (D) Hg

320. Bone flexibility disease in human being is due to high concentration of ________ is drinking water
(A) Arsenic (B) Fluorides
(C) Chlorides (D) Hg

321. Accumulation of chemicals in the tissues of biological organisms is known as
(A) Bio magnification (B) Bioaccumulation
(C) Bio transformation (D) Bio augmentation

322. E-waste (Management) rules ________
(A) 1996 (B) 2006
(C) 2016 (D) 2019

323. RDF denoted as
(A) Refuse derived fuel (B) Red degree Fahrenheit
(C) Rich diesel fuel (D) Rain derived fuel

324. BFR denoted as
(A) Brominated flame retardants
(B) Brominated fire retardants
(C) Brominated fabric retardants
(D) All

325. E waste cathode ray tubes (CRT) contains high amount of ______
(A) Phosphorus and Pb (B) Phosphorus and K
(C) Phosphorus and Hg (D) Phosphorus and Cd

326. Major E wastes generator of the world ______
(A) Ghana (B) Nigeria
(C) Guiyu, China (D) Mumbai, India

327. ISO standard related to recycling of plastics is ______
(A) ISO 15270 : 2008 (B) ISO 15270 : 2009
(C) ISO 14001 : 2004 (D) ISO 14001 : 2005

328. Without GHGs the temperature of earth would have been ______
(A) -80 °C (B) -18 °C
(C) -28 °C (D) 38 °C

329. Kyoto protocol
(A) 1977 (B) 1987
(C) 1997 (D) 2007

330. Permanent hardness in water is caused by ______
(A) Magnesium carbonate
(B) Magnesium bicarbonate
(C) Chloride and sulphate of Ca and Mg
(D) All of the above

331. Temporary hardness is caused due to ______
(A) Carbonate and Bicarbonate of Ca and Mg
(B) Calcium sulphate
(C) Magnesium sulphate
(D) Magnesium chloride

332. Permitted COD of treated effluent is ______
(A) 30 mg/l (B) 100 mg/l
(C) 250 mg/l (D) 350 mg/l

333. Which is the most polluted river in the world?
(A) Sarno river, Italy (B) Thems, England
(C) Ganges, India (D) Both A and B

334. Which is the most polluted ocean?
(A) Atlantic ocean – Gulf of Mexico
(B) Indian ocean
(C) Mediterranen sea
(D) None of the above

335. Who is responsible for safe disposal of the generated hazardous waste?
(A) Generator (B) Receiver
(C) Waste facility (D) TSDF

336. What is the goal of Hazardous Waste Management Unit?
(A) Protect health and environment by risks of HW
(B) Chemical analysis
(C) Import
(D) Export

337. Tracking of the hazardous waste is done to prevent ________
(A) Export (B) Import
(C) Midnight dumping (D) Traffic

338. Which of the following is a natural source of contamination?
(A) Organic matter (B) Petroleum
(C) Leachate (D) Septic System

339. CPCB located at
(A) New Delhi (B) Mumbai
(C) Coimbatore (D) Chennai

340. World Wide Fund for Nature is an international non-governmental organization founded in ________
(A) 1961 (B) 1971
(C) 1981 (D) 1991

341. Head quarters of World Wide Fund for Nature ________
(A) New Delhi, India (B) Gland, Switzerland
(C) Singapore (D) Tokyo, Japan

342. Head quarters of UNEP ________
(A) New Delhi, India (B) Gland, Switzerland
(C) Nairobi, Kenya (D) Tokyo, Japan

343. Head quarters of IUCN ________
(A) New Delhi, India (B) Gland, Switzerland
(C) Nairobi, Kenya (D) Tokyo, Japan

344. Friends of the Earth International ________
(A) Amsterdam, Netherlands (B) Gland, Switzerland
(C) Nairobi, Kenya (D) Tokyo, Japan

345. Conservation International head quarters is in
(A) Arlington County, Virginia, USA
(B) Gland, Switzerland
(C) Nairobi, Kenya
(D) Tokyo, Japan

346. Conservation International an American nonprofit environmental organization founded in ________
(A) 1967 (B) 1977
(C) 1987 (D) 1997

347. IPCC founded in the year ________
(A) 1968 (B) 1978
(C) 1988 (D) 1998

348. Head quarters of IPCC
(A) Arlington County, Virginia, USA
(B) Geneva, Switzerland
(C) Nairobi, Kenya
(D) Tokyo, Japan

349. Founder of IPCC
(A) Bert Bolin (B) Hoesung Lee
(C) Berlin (D) Thomson

350. Earth watch Institute headquarters ________
(A) Arlington County, Virginia, USA
(B) Boston, Massachusetts, USA
(C) Nairobi, Kenya
(D) Tokyo, Japan

351. Earth watch Institute founded in ________
(A) 1968 (B) 1978
(C) 1971 (D) 1998

352. Founder of Earth watch Institute
(A) Bert Bolin (B) Bob Citron
(C) Berlin (D) Thomson

353. Centre for Science and Environment New Delhi is founded in ______

(A) 1968 (B) 1978
(C) 1980 (D) 1998

354. Climate reality project is a non-profit organization involved in education and advocacy related to climate change is located in ______

(A) Arlington County, Virginia, USA
(B) Washington, D.C. USA
(C) Nairobi, Kenya
(D) Tokyo, Japan

355. Increase in concentration of toxic pollutant at the successive trophic level is known as ______

(A) Biomagnification (B) Fluorosis
(C) Eutrophication (D) None of these

356. High concentration of ______ disturb the calcium metabolism in birds, leads to egg shell thinning, reduction in birds population due to premature breaking

(A) Pd (B) Hg
(C) DDT (D) BHC

357. Which among the following heavy metal affects the brain and kidney?

(A) Cr (VI) (B) Cd
(C) Hg (D) Pb

358. Which among the following heavy metal is known to cause cancer?

(A) Cr (VI) (B) Cd
(C) Hg (D) Pb

359. Self purification of water by means of chemical purification is done through

(A) Calcium hydroxide (B) Calcium carbonate
(C) Calcium sulphate (D) Magnesium sulphate

360. Lime is known as ______

(A) Calcium hydroxide (B) Calcium carbonate
(C) Calcium sulphate (D) Magnesium sulphate

361. Screens and grit tanks are usually used in

(A) Primary treatment (B) Secondary treatment
(C) Tertiary treatment (D) All

362. In recent times 'ozonation' techniques is preferred for disinfection

(A) Because in chlorination techniques chloride reacts with organic compounds and form chlorinated hydro carbons which is toxic
(B) Chlorination does not provide 100% disinfection
(C) Chlorination is very expressive
(D) All the three

363. Primary treatment reduces ________ in waste water

(A) Suspended solids
(B) Oxygen demanding substances
(C) Specific pollutants
(D) All

364. Secondary treatment reduces ____________ in waste water

(A) Suspended solids
(B) Oxygen demanding substances
(C) Specific pollutants
(D) All

365. Tertiary treatment reduces ______ in waste water

(A) Suspended solids
(B) Oxygen demanding substances
(C) Specific pollutants
(D) All

366. Coarse screen used in the primary treatment consist of metal bars/ heavy wires in the space range of

(A) 10 to 20 mm (B) 25 to 50 mm
(C) 50 to 100 mm (D) 100 to 200 mm

367. The size of opening of fine screen used in the primary treatment will be around

(A) 0.8 to 6.0 mm (B) 0.8 to 6.0 cm
(C) 0.8 to 6.0 m (D) 0.8 to 6.0 m

368. Grit removal mainly removes ______

(A) Sand (B) Suspended solids
(C) BOD (D) Phosphorous

369. Skimming tank mainly removes

(A) Sand (B) Nitrate
(C) Oil and grease (D) Phosphate

370. **Which among the following is not a coagulant?**
(A) Alum (B) Ferric sulphate
(C) Ferric chloride (D) Activated silica

371. **Which among the following is added in waste water treatment to enhance coagulation process?**
(A) Alum (B) Ferric sulphate
(C) Ferric chloride (D) Activated silica

372. **Polyelectrolytes are ____________**
(A) Coagulant (B) Disinfectant
(C) Coagulant aid (D) Recalcitrant

373. **The ideal BOD : N : P ratio of waste for aerobic treatment is ________**
(A) 100 : 50 : 1 (B) 100 : 2.5 : 0.5
(C) 100 : 1 : 50 (D) 100 : 5 : 2.5

374. **The ideal BOD : N : P ratio of waste for anaerobic treatment is ________**
(A) 100: 50 : 1 (B) 100 : 2.5 : 0.5
(C) 100 : 1 : 50 (D) 100 : 5 : 2.5

375. **Formation of active suspension of biological solids in the aeration tank of waste water treatment is known as ________**
(A) Activated sludge (B) Mixed sludge
(C) Trickling filter (D) Sewage sludge

376. **The combination of activated sludge and waste water in the aeration tank is known as ________**
(A) SS - Suspended Solids
(B) TSS - Total suspended solids
(C) MLSS - Mixed liquid suspended solids
(D) TDS - Total dissolved solids

377. **The process by which primary effluent percolates downwards through the bed of stones contains biological slime of microbes is**
(A) Activated sludge process (B) Trickling filter
(C) Aerated lagoons (D) Sedimentation tank

378. **Fill, aerate, settle is a key phenomena followed in ______**
(A) Activated sludge process (B) Sequential batch reactors
(C) Tricking filter (D) Membrane Bioreactors

379. UASB means
(A) Up flow anaerobic sludge blanket
(B) Up flow aerobic sludge blanket
(C) Both A and B
(D) None of these

380. Waste water containing nitrogen can be removed by ________
(A) Sand filter (B) Carbon filter
(C) Stripping (D) Micro filtration

381. In electrolytic treatment method, quantity of silver needed for aquatic sterilization is ________
(A) 1 g of silver / 20 m^3 (B) 5 g of silver / 20 m^3
(C) 10 g of silver / 20 m^3 (A) 15 g of silver / 20 m^3

382. The concentration of solids in sewage sludge will be around ________
(A) 1 % (B) 4 %
(C) 2 % (D) 5 %

383. Activated sludge usually contains ________ solids
(A) < 1% (B) < 2%
(C) < 3% (D) < 5%

384. Sludge generated from trickling filter contains about ________ solids
(A) < 1% (B) < 2%
(C) < 3% (D) < 4%

385. Process of burning solid waste in the presence of oxygen under controlled condition is known as ________
(A) Composting (B) Open burning
(C) Pyrolysis (D) Incineration

386. Process of burning solid waste in the absence of oxygen/partial supply of oxygen then and there under controlled condition is called ________
(A) Composting (B) Open burning
(C) Pyrolysis (D) Incineration

387. When sewage water is continuously applied on land, the pores and voids in the soil gets clogged and the free circulation of air will be prevented. This process is called ________
(A) Soil sickness (B) Sewage sickness
(C) Ground water sickness (D) Sludge sickness

388. CETP means
(A) Central Executive Treatment Proposal
(B) Common Engineering Treatment Plant
(C) Common Effluent Treatment Plant
(D) Common Effluent treatment plan

389. Constructed wetlands are effective at reducing loads of BOD/COD, nitro-gen, phosphorous and suspended solids by
(A) 68% (B) 78%
(C) 78% (D) 98%

390. RZTS denotes
(A) Red zone tree system
(B) Red zone treatment system
(C) Root zone treatment system
(D) Root zone tree system

391. SFCW denotes
(A) Surface flow constructed wetlands
(B) Subsurface flow constructed wetlands
(C) Soil flow constructed wetlands
(D) Subsoil flow constructed wetlands

392. SSFC denotes
(A) Surface flow constructed wetlands
(B) Subsurface flow constructed wetlands
(C) Soil flow constructed wetlands
(D) Subsoil flow constructed wetlands

393. Which among the following is consider as free water surface constructed wetlands (FWS)
(A) SFCW (B) SSFCW
(C) RZTS (D) Vertical flow SSFCW

394. Surface flow constructed wetlands are characterized by ____________ waste water across the roots of plants
(A) Horizontal flow
(B) Vertical flow
(C) Both horizontal and vertical flow
(D) Zig zag flow

395. Which among the following is not a SSFCW treatment system?
(A) RZTS (B) VF SSFCW
(C) HF SSFCW (D) SFCW

396. The plants that grow in marsh/ wetlands (partly submerged) in water are called

(A) Halophytes (B) Psammophytes
(C) Xerophytes (D) Helophytes

397. RZTS requires pretreatment mainly to avoid

(A) Clogging (B) Surface clogging
(C) Internal clogging (D) All

398. The recommended depth of horizon-tal filters media in the SSFWC/ RZTS is

(A) 50 to 100 cm (B) 60 to 120 cm
(C) 70 to 140 cm (D) 80 to 160 cm

399. The recommended depth of vertical filters media in the SSFWC/RZTS is

(A) 50 to 100cm (B) 60 to 120 cm
(C) 70 to 140cm (D) 80 to 160 cm

400. RZTS sealing with impermeable layer is necessary at ________ to avoid water contamination

(A) the bottom alone (B) the side wall alone
(C) Both bottom and side wall (D) Sealing is not media

401. RZTS artificial sealing is not necessary for treating sewage water if the soil has a permeability coefficient of

(A) $< 10^{-8}$ m/s (B) $> 10^{-8}$ m/s
(C) $> 10^{-6}$ m/s (D) $< 10^{-6}$ m/s

402. The BOD (Organic loading) criteria range for horizontal flow RZTS will be around

(A) 10 to 20 g/BOD/m^2/day (B) 10 to 30 g/BOD/m^2/day
(C) 10 to 40 g/BOD/m^2/day (D) 10 to 50 g/BOD/m^2/day

403. The BOD (Organic loading) criteria range for vertical flow RZTS will be around

(A) 20 to 30 g/BOD/ m^2/day (B) 20 to 40 g/BOD/ m^2/day
(C) 20 to 50 g/BOD/ m^2/day (D) 20 to 60 g/BOD/ m^2/day

404. The hydraulic load criteria range for horizontal flow RZTS

(A) 200 to 1000 L/ m^2/day (B) 300 to 1000 L/ m^2/day
(C) 40 to 1000 L/ m^2/day (D) 100 to 1000 L/ m^2/day

405. The hydraulic load criteria range for vertical flow RZTS

(A) 50 to 130 L/ m^2/day (B) 500 to 1300 L/ m^2/day
(C) 5 to 13 L/ m^2/day (D) 5000 to 13000 L/ m^2/day

406. In the constructed wetlands the waste water flows horizontally over

(A) Soil substrate (B) Gravel substrate
(C) Glass substrate (D) Polyurethane foam

407. In the reed bed treatment system waste water flows vertically or horizontally over

(A) Soil substrate (B) Gravel substrate
(C) Glass substrate (D) Polyurethane foam

408. Pollutant removal process in the reed bed treatment system is

(A) Temperature dependent (B) Time dependent
(C) Gravel dependent (D) soil dependent

409. The common reed used in the reed bed treatment system is

(A) Phragmites australis (B) *Typha latifolia*
(C) *Baumea articulata* (D) *Eleocharis sphacelata*

410. The reed mace used in reed bed treatment system is

(A) *Phragmites species* (B) Typha latifolia
(C) *Barmea artiarlata* (D) *Eleocharis sphacelata*

411. CEMS denotes

(A) Continuous emission monitoring system
(B) Careful emission monitoring system
(C) Correct emission monitoring system
(D) All the tree

412. ZLD denotes

(A) Zero liquid disposal (B) Zero liquid discharge
(C) Zero liquid drain (D) Zone liquid discharge

413. Which among the following is not correct?

(A) GEMS – Genetically effective micro organisms
(B) CEMS – Continuous Emission Monitoring System
(C) CETP – Common Effluent Treatment Plant
(D) NGRSA – National Ganga River Basin Authority

414. Which among the following is not a criteria air pollutant?

(A) Ozone (B) CO
(C) NOx (D) Cd

415. it's a colourless, tasteless and odorless gas

(A) H_2S (B) CO
(C) CO_2 (D) NH_3

416. NOx represents

(A) NO and NO_3 (B) NO and N_2O

(C) NO and NO_2 (D) NO, NO_2 and NO_3

417. Reddish brown gas with pungent suffocating odour

(A) CO (B) NO

(C) NO_2 (D) H_2S

418. Which among the following metal is not present in the catalytic convertor of automobiles?

(A) Platinum (B) Palladium

(C) Rhodium (D) Lead

419. Presence of ___________ in petrol inactivate the catalysts

(A) Platinum (B) Palladium

(C) Rhodium (D) Lead

420. Aerosol consisting of solid particles is

(A) Smoke (B) Fog

(C) Smog (D) All

421. Aerosol consisting of water droplet is

(A) Smoke (B) Fog

(C) Smog (D) All

422. Reducing smog

(A) London smog (B) Los Angeles smog

(C) Bhopal smog (D) Donora smog

423. Oxidizing smog

(A) London smog (B) Los Angeles smog

(C) Bhopal smog (D) Donora smog

424. Which among the following is not photo chemical oxidant?

(A) NO_2 (B) O^3

(C) PAN (D) SO_2

425. Colourless gas with pungent odour

(A) NO_2 (B) CO

(C) SO_2 (D) NO

426. Main natural source of SO_2

(A) Forest fire (B) Volcanoes

(C) Paddy field (D) Cow dug

427. PM_{10} and $PM_{2.5}$ are fine particles with a dia

(A) 10 m and 2.5 m (B) 10 nm and 2.5 nm

(C) 10 m and 2.5 m (D) 10 cm and 2.5 cm

428. Which among the following is the major component of fly ash?
(A) Silicon (B) Potassium
(C) Iron (D) Carbon

429. Air pollutant that causes nasal irrita-tion and respiratory tract irritation are
(A) Aldehydes (B) Ammonia
(C) Arsines (D) Carbon none oxide

430. Air pollutant that damage red blood cells, damage kidney
(A) Aldehydea (B) Ammonia
(C) Arsines (D) Carbon none oxide

431. Pb $(C_2H_5)_4$ is
(A) Tetra ethyl lead (B) Tetra methyl lead
(C) Ethyl lead (D) Methyl lead

432. Which among the following is used as catalyst in the catalytic convertors to reduce NO_x and CO?
(A) Cd (B) Ni
(C) Pt (D) Th

433. Isoprene (C_3H_8) and terpenes $(C_{10}H_{15})$ examples of
(A) Halogens (B) Hydrocarbons
(C) Catalyst (D) Alcohols

434. ABC denotes
(A) American brown clouds (B) Asian brown clouds
(C) Asian black clouds (D) African brown clouds

435. Which among the following atmospheric layer is considered as global sunscreen?
(A) Troposphere (B) Stratosphere
(C) Mesosphere (D) Thermosphere

436. Ozone layer protects the life on earth by absorbing harmful UV B rays in the range of
(A) 180 to 215 nm (B) 280 to 315 nm
(C) 380 to 415 nm (D) 480 to 515 nm

437. Thinning of ozone layer over Antarctica was first observed in the year
(A) 1967 (B) 1977
(C) 1987 (D) 1997

438. Which among the following is not a ODS?
(A) CFC (B) CH_4
(C) N_2O (D) O_3

439. Unit used for the measurement of ozone thickness / concentration

(A) Bed (B) Desibel

(C) Curie (D) Dobson

440. Ozone hole is an area where O_3 concentration drops below

(A) 300 Dobson (B) 220 Dobson

(C) 100 Dobson (D) 3000 Dobson

441. Ozone thickness in the atmosphere is about

(A) 300 Dobson (B) 400 Dobson

(C) 500 Dobson (D) 600 Dobson

442. Alternatives of CFC to control ozone depletion ________

(A) Propane R290 (B) Isobutene R600A

(C) Both a and b (D) None of these

443. Scientists received Nobel price (1995) in chemistry for the work on CFC

(A) Rowland and Molina (B) Marie and Curie

(C) Roentgen (D) Odum

444. Lichens survive in highly polluted air is

(A) Bushy lichens (B) Leafy lichens

(C) Crusty lichens (D) All the three

445. Louis smog

(A) 1929 (B) 1939

(C) 1949 (D) 1959

446. Smoke pollution had been a problem in St. Louis, Missouri, USA for many decades due to

(A) Usage of high petrol

(B) Burning of bituminous

(C) Burning of wheat straw

(D) Burning of copper wires

447. Donora smog, Pennsylvania, resulted in 20 death and 7000 sickened

(A) 1946 (B) 1947

(C) 1948 (D) 1950

448. London smog/great smog of London occurred in the year ________

(A) 1952 (B) 1962

(C) 1972 (D) 1942

449. London smog killed 4000 people during
(A) October, 1952 (B) November, 1952
(C) December, 1952 (D) May, 1952

450. Bhopal gas tragedy
(A) 1st December 1984 (B) 2nd December 1984
(C) 3rd December 1984 (D) 30th November 1984

451. Compound responsible for Bhopal gas tragedy
(A) Methyl Iso Cyanate (B) Methyl Iso Cyanide
(C) Ethyl Iso Cyanate (D) Ethyl Iso Cyanide

452. ______________ and ______________ are the two compounds needed to produce MIC
(A) HCL and H_2SO_4
(B) Methylamine and Phosgene
(C) Methylamine and HCL
(D) HCL and Phosgene

453. Three mile Island disaster
(A) March 29th 1979 (B) March 29th 1989
(C) March 29th 1969 (D) March 29th 1959

454. Chernobyl nuclear disaster
(A) April 26th 1986 (B) March 29th 1979
(C) December 3rd 1984 (D) December 5th to 9th 1952

455. Thyroid cancers as a result of Chernobyl nuclear accident resulting surgical scar at the base of the neck are known as
(A) Chernobyl scar (B) Chernobyl necklace
(C) Chernobyl chair (D) Chernobyl ring

456. Forest fire that last for nearly one year (1997 to 1998) and burned over 8 million hectares of land
(A) Californian forest fire (B) Mexican forest fire
(C) Indonesian forest fire (D) Malaysian forest fire

457. Gravity settling chamber removes dust particles with size range of
(A) 25 to 30 mm (B) 5 to 10 mm
(C) <20 mm (D) 10 to 20 mm

458. Cyclone separator removes dust particles having size range of
(A) 10 to 40 mm (B) 20 to 40 mm
(C) 30 to 40 mm (D) > 40 mm

459. Method suitable for the removal of larger particles from air (dia 0.005cm)

(A) Cyclonic scrubber
(B) Venturi scrubber
(C) Electrostatic precipitators
(D) Cyclone separator

460. APTI denote

(A) Air pollution tolerance index
(B) Air pollutant table index
(C) Air pollution tolerance indice
(D) Air pollutant tolerant index

461. Which among the following is made mandatory for the highly polluting industries to install 24x7 real time monitoring of emissions and effluent discharge points?

(A) GEMS (B) CEMS
(C) APTI (D) ETP

462. CEMS means

(A) Continuous Emission Measurement System
(B) Continuous Effluent Monitoring System
(C) Continuous Emission Monitoring System
(D) Continuous Effluent Measurement System

463. Which among the following forms of chromium does not occur naturally?

(A) Cr (VI) (B) Cr (III)
(C) Cr (0) (D) All

464. Which among the following form of chromium is toxic to human beings?

(A) Cr (VI) (B) Cr (III)
(C) Cr (0) (D) All

465. Which among the following metal reduces the IQ (Intelligence Quotient) of children?

(A) Hg (B) Cr
(C) Pb (D) Ar

466. Which among the following heavy metal is present in three forms namely elements, inorganic and organic

(A) Cr (B) Hg
(C) Ni (D) Cd

467. Which among the following is used to manage contaminated soils?
(A) Air sparging (B) Soil washing
(C) Biopiles (D) All

468. The remediation of pollutant using inherent metabolic activities of micro-organisms at the polluted site is known as
(A) *Insitu* bioremediation (B) Bioventing
(C) Biosparging (D) Exsitu bioremediation

469. *In-situ* soil treatment involves supply of air and nutrients through wells to stimulate the indigenous bacteria is known as
(A) Bioventing (B) Biosparging
(C) Bioaugmentation (D) Biofilter

470. Injection of air under pressure below the ground water table to increase ground water oxygen concentration to enhance the rate of biodegradation of contaminates by native bacteria is
(A) Insitu bio remediation (B) Bioventing
(C) Biosparging (D) Exsitu bioremediation

471. Addition of indigenous / exogenous microorganisms to the contaminated site is known as
(A) Bioventing (B) Biosparging
(C) Bioaugmentation (D) Bio filter

472. Solid surfaces with microbes grown on them that are able to degrade the pollutants in the industrial air emission is
(A) Bioventing (B) Biosparging
(C) Bioaugmentation (D) Biofilter

473. Method of bioremediation in which the bacteria are motivated to start the bioremediation process is known as
(A) Bioventing (B) Bio sparging
(C) Biofilter (D) Biostimulation

474. Best suitable bioremediation method for remediating BTX pollutants are
(A) Composting (B) Bioremediation
(C) Land farming (D) Air sparging

475. BTX pollutants are
(A) Benzene, Toluene, Xylene
(B) Bromine, Toluene, Xylene
(C) Bromine, Titanium, Xylene
(D) Benzene, Titanium, Xylene

476. Use of plants to absorb, translate and store toxic contaminants from a soil matrix into their root and shoot tissue is called

(A) Phytoextraction (B) Phytostimulation
(C) Phytovolatilization (D) Phytotransformtion

477. Plants that take larger than the normal amount of contaminant from soil are known as

(A) Heavy accumulators (B) Phyto accumulators
(C) Hyper accumulators (D) Super accumulators

478. Use of rhizospheric associations between plants and symbiotic soil microbes to degrade contaminants is ________

(A) Phytoextraction (B) Rhizodegradation
(C) Phytovolatilization (D) Phytotransformation

479. Plants ability to uptake contaminants subsequently transform and volatilize the contaminants is known as

(A) Phytoextraction (B) Rhizodegradation
(C) Phytovolatilization (D) Phytotransformation

480. It is a process by which environmental contaminant gets inactivated/ degraded/ immobilized as a result of plant metabolism is

(A) Phytoextractions (B) Rhizodegradation
(C) Phytovolatilization (D) Phytotransformation

481. Vacuolar storage of contaminants in the root cells are called

(A) Phytotransformation
(B) Phytovolatilization
(C) Phytosequestration
(D) Phytostabilization

482. The word pyrolysis derived from

(A) Latin word pyro and lysis
(B) Greek word pyro and lysis
(C) French word pyro and lysis
(D) German word pyro and lysis

483. In pyrolysis pyro means

(A) Water (B) Fire
(C) Separating (D) Absence of oxygen

484. In pyrolysis 'lysis' denoted

(A) Water (B) Fire
(C) Separating (D) Absence of oxygen

485. The ideal C/N ration should be ____________ before the start composting process

(A) 10:1 (B) 30:1
(C) 40:1 (D) 50:1

486. C/N ratio during the maturity of compost will be around

(A) 14 to 18:1 (B) 28 to 30:1
(C) 38 to 40:1 (D) 8 to 10:1

487. Effective size of compostable material for quick composting is around

(A) 2 to 2.5 cm (B) 3 to 3.5 cm
(C) 4 to 4.5 cm (D) 5 to 6 cm

488. Moisture content to be maintained during composting process

(A) 20 % (B) 30%
(C) 40% (D) 60%

489. Particle size of matured good quality compost will be around

(A) < 2mm (B) 2 to 4mm
(C) 4 to 6 mm (D) 8 mm

490. Earthworm cast are popularly called as

(A) Guana (B) Black gold
(C) White gold (D) Brown gold

491. Epigeic earthworms are

(A) Surface dwellers (B) Subsurface dwellers
(C) Soil dwellers (D) Rock dwellers

492. Endogeic earthworms are

(A) Surface dwellers (B) Subsurface dwellers
(C) Soil dwellers (D) Rock dwellers

493. Anecic earthworms are

(A) Surface dwellers (B) Subsurface dwellers
(C) Soil dwellers (D) Rock dwellers

494. *Lumbricus terrestris* is a type of earthworms belong to

(A) Surface dwellers (B) Subsurface dwellers
(C) Soil dwellers (D) Rock dwellers

495. *Aporrectodea caliginosa* is a type of earthworms belong to

(A) Surface dwellers (B) Subsurface dwellers
(C) Soil dwellers (D) Rock dwellers

496. *Eisenia foetida, Eudrillus eugenia* are
   (A) Surface dwellers (B) Subsurface dwellers
   (C) Soil dwellers (D) Rock dwellers

497. Which among the following is an unpigmented earthworm?
   (A) Eisenia foatida *(B) Eudrillus eugenia*
   (C) Aporrectodea caliginosa *(D) Lumbricus terrestris*

498. *Eudrillus enginea* is also called as
   (A) African worms (B) Tiger worms
   (C) Red worms (D) Composting worms

499. *Eisenia foetida* is also called as
   (A) Tiger worms (B) Red worms
   (C) African worms (D) Both A and B

500. *Periyonyx excavates* are known as
   (A) African worms (B) Composting worms
   (C) Tiger worms (D) Red worms

Answers Key

Q. Ans	Q. Ans	Q. Ans	Q. Ans	Q. Ans	Q. Ans	Q. Ans
1. B	2. A	3. D	4. C	5. A	6. B	7. B
8. B	9. D	10. C	11. B	12. C	13. C	14. B
15. B	16. D	17. D	18. B	19. C	20. C	21. C
22. D	23. C	24. D	25. C	26. C	27. D	28. D
29. A	30. C	31. C	32. C	33. C	34. B	35. C
36. B	37. B	38. C	39. C	40. C	41. C	42. B
43. B	44. B	45. B	46. D	47. B	48. B	49. C
50. C	51. B	52. C	53. B	54. A	55. C	56. B
57. C	58. B	59. B	60. B	61. C	62. D	63. B
64. A	65. B	66. A	67. C	68. C	69. A	70. A
71. B	72. C	73. B	74. D	75. D	76. A	77. D
78. B	79. A	80. C	81. B	82. B	83. A	84. C
85. A	86. B	87. B	88. B	89. C	90. B	91. D
92. C	93. B	94. C	95. A	96. D	97. B	98. B
99. B	100. C	101. C	102. B	103. C	104. C	105. D
106. C	107. C	108. C	109. C	110. B	111. A	112. A
113. D	114. C	115. A	116. A	117. B	118. B	119. C
120. D	121. A	122. C	123. A	124. A	125. B	126. A
127. B	128. A	129. D	130. B	131. B	132. C	133. B
134. A	135. D	136. C	137. C	138. C	139. B	140. C

Q. Ans	Q. Ans	Q. Ans	Q. Ans	Q. Ans	Q. Ans	Q. Ans
141. A	142. D	143. B	144. D	145. A	146. C	147. C
148. B	149. B	150. B	151. A	152. B	153. B	154. D
155. A	156. C	157. B	158. D	159. D	160. A	161. C
162. B	163. A	164. D	165. B	166. C	167. C	168. B
169. B	170. A	171. C	172. C	173. B	174. A	175. B
176. C	177. A	178. B	179. C	180. A	181. C	182. D
183. A	184. D	185. D	186. A	187. A	188. A	189. D
190. D	191. A	192. A	193. D	194. D	195. D	196. A
197. D	198. D	199. B	200. C	201. B	202. C	203. A
204. C	205. D	206. D	207. B	208. D	209. D	210. B
211. C	212. C	213. A	214. D	215. D	216. C	217. C
218. D	219. B	220. B	221. A	222. A	223. C	224. B
225. A	226. C	227. D	228. B	229. C	230. C	231. C
232. D	233. A	234. D	235. C	236. A	237. C	238. C
239. D	240. B	241. C	242. C	243. A	244. D	245. B
246. A	247. A	248. A	249. A	250. A	251. A	252. B
253. B	254. C	255. C	256. A	257. B	258. B	259. A
260. A	261. D	262. D	263. A	264. C	265. B	266. B
267. D	268. B	269. D	270. A	271. C	272. C	273. D
274. C	275. D	276. B	277. A	278. A	279. A	280. C
281. C	282. A	283. B	284. B	285. C	286. A	287. A
288. A	289. A	290. B	291. B	292. A	293. A	294. D
295. A	296. A	297. B	298. B	299. D	300. D	301. B
302. A	303. B	304. A	305. A	306. C	307. C	308. C
309. A	310. B	311. D	312. A	313. C	314. C	315. B
316. C	317. B	318. D	319. B	320. B	321. B	322. C
323. A	324. A	325. A	326. C	327. A	328. B	329. C
330. C	331. A	332. C	333. C	334. A	335. A	336. A
337. C	338. A	339. A	340. A	341. B	342. C	343. B
344. A	345. A	346. C	347. C	348. B	349. A	350. B
351. C	352. B	353. C	354. B	355. A	356. C	357. D
358. A	359. B	360. B	361. A	362. A	363. A	364. B
365. C	366. B	367. A	368. A	369. C	370. D	371. D
372. C	373. A	374. B	375. A	376. C	377. B	378. B
379. A	380. C	381. A	382. D	383. A	384. B	385. D
386. C	387. B	388. C	389. D	390. C	391. A	392. B
393. A	394. A	395. D	396. D	397. D	398. A	399. B
400. C	401. A	402. B	403. B	404. C	405. A	406. A
407. B	408. B	409. A	410. B	411. A	412. B	413. A
414. D	415. B	416. C	417. C	418. D	419. D	420. A

Q. Ans	Q. Ans	Q. Ans	Q. Ans	Q. Ans	Q. Ans	Q. Ans
421. B	422. A	423. B	424. D	425. C	426. B	427. A
428. A	429. A	430. C	431. A	432. C	433. B	434. B
435. B	436. B	437. C	438. D	439. D	440. B	441. A
442. C	443. A	444. C	445. B	446. B	447. C	448. A
449. C	450. C	451. B	452. B	453. A	454. A	455. B
456. C	457. A	458. A	459. C	460. A	461. B	462. C
463. C	464. A	465. C	466. B	467. A	468. D	469. A
470. B	471. C	472. D	473. D	474. C	475. A	476. A
477. C	478. B	479. C	480. D	481. C	482. B	483. B
484. C	485. B	486. A	487. A	488. D	489. A	490. B
491. A	492. B	493. C	494. C	495. B	496. A	497. C
498. A	499. D	500. B				

◈

CHAPTER 2

AIR POLLUTION

1. An average human being inhales about ________ litres of air every day.
   (A) 11,000 (B) 1000
   (C) 100 (D) 10

2. Air (Prevention and Control of Pollution) Act was enacted in the year ________
   (A) 1980 (B) 1981
   (C) 1982 (D) 1983

3. Air (Prevention and Control of Pollution) Rules was enacted during ________
   (A) 1980 (B) 1981
   (C) 1982 (D) 1983

4. The concentration of ozone in dry and clean air is about ______ % by volume.
   (A) 0.000002 (B) 0.00004
   (C) 0.0002 (D) 0.000004

5. Reason for the accumulation of CFCs in the atmosphere? Find out the correct statement.
   1. Escape into atmosphere during use
   2. Scrapping of refrigerators & air conditioners
   3. Disposal of foam plastics

   (A) 1 & 2 (B) 2 & 3
   (C) 2 only (D) 1, 2 & 3

6. Distance of Ozone layer from the sea level is ________ Km
   (A) 12 (B) 13 (C) 14 (D) 15

7. Example for second hand smoking is ________
   (A) Cigarette (B) Burning of fossil fuels
   (C) Industrial smoke (D) Smoke from fire

8. **Unpolluted rain water is slightly acidic owing to the presence of ___________ in the air.**
   (A) O_2 (B) NO_2
   (C) SO_2 (D) CO_2
9. **Rain water with pH lower than ____ is called acid rain.**
   (A) 5.7 (B) 5.0
   (C) 6.7 (D) 7.0
10. **Colour of Ozone gas is ______**
   (A) Pale blue (B) Pale yellow
   (C) Pale green (D) Pale white
11. **Thinning of Ozone layer may lead to an increase of ______**
   (A) Skin cancer (B) Eye cataract
   (C) (A) & (B) (D) (A) alone
12. **Berylliosis is caused by ______**
   (A) Mercury (B) Lead
   (C) Cadmium (D) Beryllium
13. **Arsenic cause cancer of ______**
   (A) Skin (B) Lungs
   (C) Kidney (D) (A) & (B)
14. **Methane gas comprises _______ % by volume of the air.**
   (A) 0.00022 (B) 0.0044
   (C) 0.0022 (D) 0.00044
15. **The most obvious damage caused by air pollution to vegetation occurs in _______**
   (A) Leaf structure (B) Tip
   (C) Stem (D) Crop yield
16. **Lichens are very sensitive to ______**
   (A) O_2 (B) NO_2
   (C) SO_2 (D) CO_2
17. **Tip burning of white pine seedlings are caused by ______**
   (A) O_3 (B) NO_2
   (C) SO_2 (D) CO_2
18. **In small concentrations, ___________ may cause dizziness and headache.**
   (A) O_2 (B) NO_2
   (C) SO_2 (D) CO

19. Causes irritation of the respiratory system
    (A) O_2 (B) NO_2
    (C) SO_2 (D) CO_2

20. Dermatitis is caused by ________
    (A) Hydrocarbons (B) Copper
    (C) Mercury (D) Arsenic

21. The inhalation of asbestos may result in ________
    (A) Bronchial carcinoma (B) Mesothelioma
    (C) B alone (D) (A) & (B)

22. ______ can cause damage to Central Nervous System and death
    (A) Boron (B) Boron hydrides
    (C) Methyl bromide (D) Sodium bromide

23. Chlorine is a disinfectant causes ________
    (A) Skin irritation (B) Eye irritation
    (C) Throat infection (D) Cancer

24. Sleeplessness is caused by ______
    (A) H_2S (B) Hg
    (C) Cd (D) Fe

25. Lead encephalopathy is encountered more in______
    (A) Adults (B) Children's
    (C) Young's (D) Old age

26. Which gas is mainly produced due to incomplete burning of wood?
    (A) CO (B) SO_2 (C) NO_2 (D) NO_3

27. The maximum size of fly ash is
    (A) 1μm (B) 100μm (C) 1000μm (D) 10μm

28. Which of the following is a secondary air pollutant?
    (A) SPM (B) PAN
    (C) SO_2 (D) NO_2

29. Which air pollutant cause corrosion of building?
    (A) SO_2 (B) SO_3
    (C) CO (D) NO_2

30. Which of the following is not a part of photochemical smog?
    (A) NO_2 (B) O_3
    (C) PAN (D) SPM

31. **The major contributor of Carbon monoxide is**
(A) Motor vehicle
(B) Industrial processes
(C) Stationary fuel combustion
(D) None

32. **Ozone is formed in the upper atmosphere by a photochemical reaction with**
(A) Ultra violet solar radiation (B) Infra-red radiation
(C) Visible light (D) All of the above

33. **Bhopal gas tragedy was due to leakage of ________ gas**
(A) Methyl Cyanide (B) Methyl ISO Cyanate
(C) Methyl amine (D) Phosgene

34. **Ozone of found in**
(A) Mesosphere (B) Ionosphere
(C) Stratosphere (D) Exosphere

35. **Which of the following is the best indicator of SO_2 pollution?**
(A) Bryophyte (B) Pteridophyte
(C) Lichens (D) Algae

36. **Cement factory labourers are prone to**
(A) Leukemia (B) Asbestosis
(C) Cytosilicosis (D) Bone marrow disease

37. **Greenhouse effect was discovered by________**
(A) de Voeux (B) Charles Dufay
(C) Joseph Fourier (D) John Gorton

38. **World Ozone Day is celebrated on ________**
(A) 16 June (B) 16 July
(C) 16 August (D) 16 September

39. **The toxic form of mercury is________**
(A) Oxidized form (B) Methyl mercury
(C) Mercuric acid (D) Elementary Mercury

40. **The major photochemical oxidant is:**
(A) Ozone (B) Hydrogen peroxide
(C) Nitrogen oxides (D) Peroxyl Acetyl Nitrate (PAN)

41. **Which of the following devices is suitable for the removal of gaseous pollutants?**
(A) Cyclone separator (B) Electrostatic precipitator
(C) Fabric filter (D) Wet scrubber

42. Among the following, the only secondary pollutant is ______
(A) SO_2 (B) O_3
(C) SO_4 (D) SO_3

43. Which of the following on inhalation dissolved in the blood hemoglobin more rapidly than oxygen?
(A) SO_2 (B) CO
(C) SO_4 (D) NO_3

44. Which of the following are likely to be present in photochemical smog?
(A) Sulphur dioxide (B) Photochemical oxidants
(C) Chlorofluorocarbon (D) Smog

45. Air pollution from automobiles can be controlled by fitting
(A) Cyclone separator (B) Electrostatic precipitator
(C) Catalytic separator (D) Wet scrubber

46. Taj Mahal at Agra may be damaged by
(A) Sulphur dioxide (B) Chlorine
(C) Hydrogen (D) Oxygen

47. Which of the following is used as antiknock compound in gasoline?
(A) Tetramethyllead (B) tetraethyl lead
(C) trimethyl lead (D) triethyl lead

48. The boiler flue gas is source of ______
(A) HCl (B) NO
(C) HF (D) Volatile Organic Compounds

49. All are particulate pollutants except
(A) Dust (B) Ozone
(C) Soot (D) Smoke

50. The supersonic jets cause pollution by the thinning of ______ layer
(A) CO_2 (B) SO_2 (C) O_3 (D) O_2

51. What are the sources of natural air pollutants?
(A) Volcanoes (B) Forest fire
(C) Wind storms (D) All

52. The satellite used to measure the Ozone thickness of Antarctica region is ______
(A) Nimbus 4 (B) Nimbus 7
(C) Nimbus 8 (D) Nimbus 3

53. Volcanoes form the major pollutant is ________
(A) Sulphur oxide (B) Carbon dioxide
(C) Nitrogen oxide (D) Hydrocarbon

54. Forest fire produce pollution through ________
(A) Carbon monoxide (B) Carbon dioxide
(C) Nitrogen oxide (D) All

55. Air is mixture of gases, primarily ________________ (78%) and ________________ (21%)
(A) Oxygen, Nitrogen (B) Carbon dioxide, Oxygen
(C) Nitrogen, Carbon dioxide (D) Nitrogen, Oxygen

56. Air we breathe is ______ resource that is replenished by natural processes. For example Oxygen is replenished by plants.
(A) Nonrenewable (B) Renewable
(C) Transparent (D) Natural

57. The pollutants in atmosphere arise from ________ sources
(A) Natural (B) Artificial
(C) Anthropogenic (D) Option (A) and (C)

58. Which natural event produces huge quantities of air pollution each day ________
(A) Volcanoes (B) Dust storms
(C) Human activities (D) Forest fire

59. The criteria air pollutants are generated by three principal sources ________
i. Transportation
ii. Various Industrial Process
iii. Fuel combustion at stationary sources
(A) Only i (B) i and ii (C) All above (D) i and iii

60. The federal ________ was our nation's first real attempt to reduce air pollution
(A) Clean Air Act
(B) Central Pollution Control Board
(C) Water Pollution Control Act
(D) Air Rule, 1982

61. ___________ that prevents lethal amounts of ultraviolet from reaching the Earth's surface.
(A) Ionosphere (B) Ozone layer
(C) Troposphere (D) Thermosphere

62. These region in which high-energy charged particles travelling outwards from the Sun have been captured by the Earth's magnetic field ____________

(A) Aurora australis (B) Aurora borealis
(C) Van Allen Radiation belt (D) Solar wind

63. The measurement of noise is given by the scientist ________

(A) David Beckham (B) Alfred Graham Bell
(C) Alexander Graham Bell (D) Michael Faraday

64. The atmosphere selectively absorbs ______ solar radiations at an altitude of 50 km or above it.

(A) UV A (B) UV B
(C) UV A & UV B (D) UV C

65. Indian Air Amendment Act _______

(A) 1987 (B) 1980
(C) 1976 (D) 1970

66. With limited supply of oxygen, the exhaust may contain certain amount of carbon monoxide and carbon particles known as soot. This is due to

(A) Incomplete combustion (B) Combustion
(C) Reduction (D) Oxidation

67. Greenhouse effect will cause _______

i) Expansion of ocean ii) Gradual melting of polar ice caps iii) Changing rainfall pattern

(A) Only i (B) Option ii & iii
(C) Option i & ii (D) All the above

68. How CFCs are accumulated?

(A) Escape into the atmosphere during use
(B) Scrapping of refrigerator and air conditioners
(C) Disposal of foam plastics (D) All

69. ____________ is famous for its episode of industrial smog. (December 1952)

(A) New York (B) Los Angeles
(C) London (D) Sydney

70. Things you can do for cleaner air __________

(A) Use cruise control on the highway
(B) Don't top off at the gas pump
(C) Choose products with recyclable packaging
(D) All

71. The effective methods to Control Air Pollution are as follows:

(A) Source Correction Methods
(B) Pollution Control equipment
(C) Vegetation and Zoning
(D) All

72. ______ is a measurement of column ozone level

(A) Dobson level (B) Decibel
(C) Km (D) Ft

73. ______ is a term used internationally to describe air pollutants that have been regulated and are used as indicators of air quality?

(A) Primary pollutants (B) Criteria of air pollutants
(C) Air pollution (D) Pollution

74. Primary criteria pollutants include the gases ______

I. Sulphur dioxide
II. Nitrogen oxide
III. Carbon monoxide
IV. Solid or liquid particulates

(A) Option I, II and III (B) Option I and II
(C) Option III (D) All

75. The term smog was coined by ______

(A) Harrold A. Desvaeux (B) Tansley
(C) Harrold (D) Odum

76. Cigarette smoke contains more than ______ of carbon monoxide

(A) 200 ppm (B) 150 ppm
(C) 400 ppm (D) 100 ppm

77. Which of the following can cause depletion of ozone?

(A) H_2S (B) NO
(C) Smoke (D) Aerosols

78. Colourless, odourless & non-corrosive air pollutant is ______

(A) SO_2 (B) CO
(C) CO_2 (D) O_3

79. The presence of which gas can check the UV light from sunlight?

(A) SO_2 (B) NO
(C) CO_2 (D) O_3

80. Mercury is emitted into air by ________

(A) Burning woods (B) burning garbage

(C) Coal fire (D) Steam engine

81. Which of the following causes warming of atmosphere but cooling of the earth's surface ?

(A) Ozone (B) Black carbon aerosols

(C) All greenhouse gases (D) Sulphates & nitrates

82. Which of the air pollutants reacts with lead-based paints causing discolouration?

(A) SO_2 (B) H_2SO_4

(C) HNO_3 (D) H_2S

83. The Pollution Standard Index (PSI) scale has span from ________

(A) 0-200 (B) 0-300

(C) 0-400 (D) 0-500

84. Which of the following is an organic gas?

(A) Hydrocarbons (B) Aldehydes

(C) Ketones (D) Ammonia

85. Which of the following is/are inorganic gas(es)?

(A) Carbon monoxide (B) Hydrogen sulphide

(C) Chlorine (D) All of the above

86. The major contributor of Carbon monoxide is ______

(A) Motor vehicle

(B) Industrial processes

(C) Stationary fuel combustion

(D) None of the above

87. Hypoxia ______

(A) CO (B) NO

(C) SO (D) SO_2

88. Fugitive emissions consist of ______

(A) Street dust

(B) Dust from construction activities

(C) Dust from farm cultivation

(D) All of the above

89. Ozone of found in ________

(A) Mesosphere (B) Ionosphere

(C) Stratosphere (D) Exosphere

90. **The principal source of volatile organics (Hydrocarbons) is ______**

(A) Transportation
(B) Industrial processes
(C) Stationary fuel combustion
(D) Volcanoes

91. **The function of automobile catalytic converter is to control emissions of ______**

(A) Carbon dioxide and hydrogen
(B) Carbon monoxide and hydrogen
(C) Carbon monoxide and carbon dioxide
(D) Carbon monoxide and nitrogen dioxide

92. **The list of industrial sources of air pollution and their emissions are given. Match the following.**

A. Ammonia	1. Carbon monoxide
B. Plating	2. Particulates
C. Fertilizers	3. Metal fumes

The correct order is

(A) A-1, B-2, C-3 (B) A-3, B-2, C-1
(C) A-1, B-3, C-2 (D) A-3, B-1, C-2

93. Air pollution causes ______

(A) Global warming (B) Respiratory problems
(C) Soil erosion (D) None of these

94. Green House gas is ______

(A) Nitrogen (B) Oxygen
(C) Methane (D) Carbon dioxide

95. The percentage of nitrogen in the atmosphere is ______

(A) 21% (B) 78 %
(C) 12% (D) 87%

96. Chlorofluorocarbon is used in ______

(A) Refrigerators (B) Air conditioners
(C) Perfumes (D) All

97. Which of the following diseases are caused by the smog?(i) Rickets, (ii) Throat Cancer, (iii) Skin Cancer, (iv) Breathing Problem

(A) both (i) and (ii) (B) both (iii) and (iv)
(C) (i), (ii) and (iii) (D) (i), (ii) and (iv)

98. **Which country release maximum green house gases?**

(A) India (B) USA

(C) China (D) Russia

99. **1 Dobson Unit at std temperature is equal to ozone column thickness of ______**

(A) 0.1 mm (B) 0.01 mm

(C) 0.001 mm (D) 1 mm

100. **The disease caused by cotton fiber in air ______**

(A) Byssinosis (B) Bronchitis

(C) Berylliosis (D) Anemia

Answers Key

Q. Ans	Q. Ans	Q. Ans	Q. Ans	Q. Ans	Q. Ans	Q. Ans
1. A	2. B	3. C	4. A	5. D	6. D	7. A
8. D	9. A	10. A	11. C	12. D	13. D	14. A
15. A	16. C	17. A	18. D	19. C	20. B	21. D
22. B	23. B	24. A	25. B	26. A	27. C	28. B
29. A	30. D	31. A	32. A	33. B	34. C	35. C
36. C	37. C	38. D	39. B	40. C	41. D	42. B
43. B	44. C	45. C	46. A	47. B	48. B	49. B
50. C	51. D	52. A	53. A	54. D	55. D	56. B
57. D	58. C	59. C	60. A	61. B	62. C	63. B
64. B	65. A	66. A	67. D	68. D	69. C	70. D
71. D	72. A	73. C	74. D	75. A	76. C	77. D
78. B	79. D	80. B	81. B	82. D	83. D	84. D
85. D	86. A	87. A	88. D	89. C	90. C	91. B
92. C	93. B	94. D	95. B	96. D	97. D	98. B
99. B	100. A					

CHAPTER 3

ATMOSPHERIC CHEMISTRY

1. ________ is a branch of atmospheric science in which the chemistry of the Earth's atmosphere and that of other planets is studied.
   (A) Atmospheric chemistry (B) Atmospheric physics
   (C) Meteorology (D) All the above.

2. ________ is a colourless, odourless and tasteless physical mixture of gases which surrounds earth on all sides.
   (A) Atmosphere (B) Biosphere
   (C) Lithosphere (D) Hydrosphere

3. The lower atmosphere where the chemical composition of gas is uniform is called ______
   (A) Atmosphere (B) Biosphere
   (C) Lithosphere (D) Homosphere

4. At higher levels the chemical composition of air changes considerably and known as ______
   (A) Atmosphere (B) Heterosphere
   (C) Lithosphere (D) Homosphere

5. Average composition of atmosphere ______
   (A) Nitrogen = 74%, Oxygen = 25%, Carbon dioxide=0.03% and Rest of Others
   (B) Nitrogen = 78%, Oxygen =0.03%, Carbon dioxide=21% and Rest of Others
   (C) Nitrogen = 70%, Oxygen = 21%, Carbon dioxide=0.3% and Rest of Others
   (D) Nitrogen = 78%, Oxygen = 21%, Carbon dioxide=0.03% and Rest of Others

6. Coldest layer of the atmosphere is ______
   (A) Stratosphere (B) Mesosphere
   (C) Thermosphere (D) Exosphere

7. **Which atmospheric layer is called as Seat of photochemical reaction?**
(A) Stratosphere (B) Mesosphere
(C) Thermosphere (D) Exosphere

8. **Which atmospheric layer is called as Seat of weather phenomenon?**
(A) Stratosphere (B) Mesosphere
(C) Troposphere (D) Exosphere

9. **Radio communication occurs in which layer of atmosphere?**
(A) Stratosphere (B) Mesosphere
(C) Thermosphere (D) Exosphere

10. **The decrease in air temperature with height is known as the________**
(A) Reduce rate (B) Lapse rate
(C) Decreasing rate (D) Normal rate

11. **The lapse rate of our atmosphere is ________**
(A) 5.6 °C/km (B) 6.5 °C/km
(C) 8.5 °C/km (D) 5.8 °C/km

12. **Which gases is predominant in exosphere layer of atmosphere?**
(A) H and CO_2 (B) H_2 and He
(C) H and He (D) O_2 and CO_2

13. **Who discovered the ozonosphere?**
(A) Joseph Priestley
(B) Antoine Lavoisier
(C) Christian Friedrich Schoenlein
(D) Henry Cavendish

14. **Which layer of the atmosphere has the highest density of gas molecules?**
(A) Exosphere
(B) Mesosphere
(C) Stratosphere
(D) Troposphere

15. **In which layer do auroras (*e.g.* northern lights) occur?**
(A) Mesosphere
(B) Stratosphere
(C) Thermosphere
(D) Troposphere

16. **Which two atmospheric layers have temperature profiles that promote convection?**
(A) Mesosphere and Stratosphere
(B) Mesosphere and Thermosphere
(C) Mesosphere and Troposphere
(D) Stratosphere and Thermosphere

17. **What frequencies of electromagnetic radiation are most abundantly emitted by the earth?**
(A) Infrared light (B) Microwaves
(C) Radio waves (D) Ultraviolet light

18. **__________ is a part of a routine climate pattern that occurs when sea surface temperatures in the tropical Pacific Ocean rise to above-normal levels for an extended period of time.**
(A) Na nino (B) El nino
(C) Hurricane (D) Typhoon

19. **El Niño is a part of a routine climate pattern that occurs when sea surface temperatures in the __________ rise to above-normal levels for an extended period of time.**
(A) Northern Atlantic Ocean (B) Arctic Ocean
(C) Tropical Pacific Ocean (D) Indian Ocean

20. **The warm and cool phases are part of a recurring climate pattern that occurs across this section of the Pacific, known as the __________**
(A) ENSO (B) NNSO
(C) TPSO (D) NOAA

21. **ENSO stands for ________**
(A) El Nino-Southern Oscillation
(B) European Southern Oscillation
(C) El Nino Southern Organization
(D) Environmental Southern Oscillation

22. **What is the percentage of argon in atmosphere?**
(A) 0.90% (B) 0.82%
(C) 0.93% (D) 0.99%

23. **The layer between the troposphere and stratosphere is __________**
(A) Tropopause (B) Stratopause
(C) Mesopause (D) Thermosphere

24. Chemosphere is otherwise called as ________

(A) Troposphere (B) Stratosphere
(C) Mesosphere (D) Ionosphere

25. ________ is the study of the Earth's atmosphere, its processes, the effects other systems have on the atmosphere, and the effects of the atmosphere on these other systems.

(A) Meteorology (B) Atmospheric science
(C) Earth science (D) Atmospheric chemistry

26. ________ is the meteorological science of the upper region of the Earth's or other planetary atmosphere.

(A) Aerology (B) Aeronomy
(C) Aerosphere (D) All the above

27. ________ is the weight of the air, which lies vertically above a unit areacentered at a point.

(A) Air temperature (B) Atmospheric pressure
(C) Atmospheric effect (D) Earth's pressure

28. Air is deflected towards right in the Northern Hemisphere and Left in the Southern hemisphere due to rotation of earth and this phenomenon is called

(A) Newton's force (B) Coriolis force
(C) Gravitational force (D) Rotation force

29. ________ is when sea surface temperatures in the central Pacific drop to lower-than-normal levels.

(A) La Nina (B) El Niño
(C) Hurricane (D) Typhoon

30. El Niño occurs in which continent?

(A) North America (B) South America
(C) Australia (D) Europe

31. ENSO cycle includes ________

(A) La Nino (B) El Niño
(C) Both A and B (D) None of the above

32. According to the ENSO, El Niño and La Niño occurs in which phase?

(A) Cool and warm phase (B) Warm and cool phase
(C) Cool phase (D) Warm phase

33. ________ applied to an annual weak warm ocean current that ran southwards along the coast of Peru and Ecuador at about Christmas time.

(A) La Nina (B) El Niño
(C) Both A and B (D) None of the above

34. Strongest record of el Niño occurs in which year?
(A) 2017-2019 (B) 2013-2015
(C) 2014-2016 (D) 2010-2012

35. When this warming occurs for seven to nine months, it is classified as El Niño "conditions"; when its duration is longer, it is classified as an ________
(A) La Nino episode (B) El Niño episode
(C) Both A and B (D) None of the above

36. Who coined the term "Southern Oscillation"?
(A) Gilbert Walker
(B) Antoine Lavoisier
(C) Christian Friedrich Schoenlein
(D) Henry Cavendish

37. The traditional Niño, also called as ________
(A) Northern Pacific (NP) El Niño
(B) Southern Pacific (SP) El Niño
(C) Western Pacific (WP) El Niño
(D) Eastern Pacific (EP) El Niño

38. La Niño occurs in which continent?
(A) North America (B) South America
(C) Australia (D) Europe

39. The ________ also contains 99% of the water vapor in the atmosphere.
(A) Mesosphere (B) Stratosphere
(C) Thermosphere (D) Troposphere

40. Average height of the troposphere from the MSL (Mean Sea Level) is ____
(A) 15 km (B) 14 km (C) 10 km (D) 19 km

41. In which layer, about 75 % of total gases and most of the moisture and dust particles present.
(A) Mesosphere (B) Stratosphere
(C) Thermosphere (D) Troposphere

42. Less convection takes place in which layer?
(A) Mesosphere (B) Stratosphere
(C) Thermosphere (D) Troposphere

43. Thermodynamic transformation which occurs without exchange of heat between a system and its environment is known as
(A) Thermodynamic process
(B) Adiabatic process
(C) Diabatic process
(D) Hemodynamic process

44. Adiabatic cooling accompanies in which process?
(A) Compression (B) Expansion
(C) Conduction (D) Convection

45. __________ involves the study of observations and theory dealing with all motion systems of meteorological importance.
(A) Atmospheric chemistry
(B) Atmospheric physics
(C) Meteorology
(D) Atmospheric dynamics

46. __________ is the study of atmospheric changes (both long and short-term) that define average climates and their change over time, due to both natural and anthropogenic climate variability.
(A) Atmospheric chemistry (B) Atmospheric physics
(C) Meteorology (D) Climatology

47. When the air rises, either because warmer or because of the terrain is called a
(A) Conduction (B) Compression
(C) Expansion (D) Convection

48. The separation between polar cold air and temperature mild air is called a
(A) Polar solvent (B) Polar front
(C) Polar back (D) Polar strip

49. Clouds formed in layers from an inversion is called as __________
(A) Cumuliform (B) Stratiform
(C) Cirriform (D) None of the above

50. __________ is the envelope of gases surrounding the earth or another planet.
(A) Atmosphere (B) Biosphere
(C) Lithosphere (D) Hydrosphere

Answers Key

Q. Ans	Q. Ans	Q. Ans	Q. Ans	Q. Ans	Q. Ans	Q. Ans
1. A	2. A	3. D	4. B	5. D	6. B	7. A
8. C	9. C	10. B	11. B	12. B	13. C	14. D
15. C	16. C	17. A	18. B	19. C	20. A	21. A
22. C	23. A	24. C	25. B	26. B	27. B	28. B
29. A	30. B	31. C	32. B	33. B	34. C	35. B
36. A	37. D	38. A	39. D	40. B	41. D	42. B
43. B	44. B	45. D	46. D	47. D	48. B	49. B
50. A						

CHAPTER 4

SOIL POLLUTION

1. Contamination of soil system by considerable quantities of chemicals and substance is known as ________

   (A) Water pollution (B) Soil Pollution
   (C) Air pollution (D) Noise Pollution

2. Soil Pollution as part of land degradation is caused by the presence of ________ (human made) chemicals or other alteration in the natural soil environment

   (A) Zenobiotic (B) Xenobiotic
   (C) Senobiotic (D) Xenobiosis

3. Adding or dumping of ________ in soil causes soil pollution

   (A) Mercury (B) Pollutants
   (C) Pesticides (D) All

4. Types of soil pollutant present in the earth ________

   (A) 2 (B) 4 (C) 7 (D) 5

5. ________ work inside the soil to release nutrients

   (A) Biological agents (B) Chemical agents
   (C) Urban wastes (D) Industrial wastes

6. Which among the following are soil pollutants?

   (A) Lead (B) Mercury
   (C) Arsenic (D) All the above

7. Which of the following is NOT considered as pollutant?

   (A) Chemical pollutant (B) Trash
   (C) Fertilizer (D) Compost

8. Which of the following is NOT an effect of soil pollution?

   (A) Severe illness in living things
   (B) Plants grow in soil filled with toxic chemicals
   (C) Oceans get polluted
   (D) New trees will grow healthy and strong

9. **Which of the following waste consists of garbage, rubbish materials, dried sludge and sewage of domestic and commercial origin?**

(A) Rural waste (B) Urban waste
(C) Industrial waste (D) Radioactive waste

10. **Substances such as Radium, Thorium, Uranium, Nitrogen, etc. can infiltrate the soil and create toxic effects and these are**

(A) Radioactive material (B) Chemical material
(C) Agricultural material (D) Medical material

11. **Steel, pesticides, textiles, drugs, glass, cement, petroleum, etc. are considered as ________**

(A) Rural waste (B) Urban waste
(C) Industrial waste (D) Radioactive waste

12. **Soil polluted by pesticides, fertilizers, herbicides, slurry, debris, and manure is called ________**

(A) Radioactive pollution (B) Chemical pollution
(C) Agricultural pollution (D) Medical pollution

13. **Which among the following chemical is a potential source of mining, coal burning, alkali and metal processing industry?**

(A) Lead (B) Mercury
(C) Arsenic (D) Cadmium

14. What is PAHS?

(A) Poly Aromatic Hydrocarbons
(B) B. Poly Aliphatic Hydrocarbons
(C) Polycyclic Aromatic Hydrocarbons
(D) D. Polycyclic Aliphatic Hydrocarbons

15. Which one of the following is not a major source of soil pollution?

(A) Coal mining
(B) Coal based thermal power plants
(C) Land filling of municipal wastes
(D) Textile units

16. An example of PAH compound is ________

(A) PAN (B) TEL
(C) BHC (D) Benzopyrene

17. **In one year, how many tonnes of top soils are lost across the Globe?**

(A) 24 billons (B) 20 billons
(C) 30 billons (D) 25 billons

18. The continued capacity of the soil to function as a vital living system, within ecosystem and land-use boundaries, to sustain biological productivity, promote the quality of air and water environments, and maintain plant, animal, and human health is ________.

    (A) Soil contamination (B) Soil health
    (C) Soil pollution (D) Soil fertility

19. ______ occurs when the concentration of a chemical or substance is higher than it would occur naturally but is not necessarily causing harm.

    (A) Soil contamination (B) Soil health
    (C) Soil pollution (D) Soil fertility

20. Which continent faces high soil pollution presently?

    (A) Australia (B) Asia
    (C) Europe (D) America

21. Which of the following is the major point source of soil pollution?

    (A) Natural phenomenon (B) Anthropogenic activities
    (C) Chemical activities (D) Environmental changes

22. In India, which pollutant cause more effect in agricultural fields?

    (A) Chromium and nickel (B) Arsenic and lead
    (C) Mercury and lead (D) Lead and cadmium

23. Soil pollutants are

    (A) Organic Pollutant
    (B) Inorganic Pollutants
    (C) Both organic and inorganic pollutants
    (D) None of these

24. Mitti Bachao Abhiyan movement launched in the year 1977 aims at ________

    (A) Save soil (B) Deforestation
    (C) Air protection (D) Water protection

25. Mitti Bachao Abhiyan movement was launched against water logging and soil salinity caused by Tawa dam of

    (A) MP, India (B) UP, India
    (C) North Africa (D) TN, India

26. Increase of 1 °C in air temperature produces an increase in the volatilization of POPs from soils by what percentage?
(A) 9% (B) 8%
(C) 6% (D) 5%

27. Which of the following plant control the soil pollution?
(A) Castor (B) Soybean
(C) Sorghum (D) Pungam

28. Why castor bean plant fighting against soil pollution?
(A) It absorb heavy metals from soil
(B) It reduce the microorganism activities in soil
(C) It absorbs more soil component from the soil
(D) It changes the soil physical properties.

29. Which of these factors is not a cause of increase in land pollution?
(A) Increase in solid waste of garbage on open lots
(B) Toxic waste
(C) Proper disposal of cans in nearby bins
(D) Radioactive waste

30. Solutions for mitigating soil pollution
(A) Proper waste disposal that focuses on treating waste and disposing it in the safest manner possible.
(B) Reusing materials to reduce the need for harvesting of resources. Products that are not reusable can likely be recycled.
(C) Organic gardening can reduce the usage of pesticides and insecticides. Non-gardeners can help by buying organic food.
(D) All the above

31. Which of the following is not an effect of land pollution?
(A) Ground water poisoning
(B) Loss of topsoil
(C) Shifting habitat
(D) Combustion of fossil fuels

32. How can land pollution harm the environment?
(A) It is ugly and can destroy the beauty of nature
(B) It can destroy the habitats of animals
(C) It can allow harmful chemicals to enter the soil
(D) All the above

33. Collection of organic waste and breaks down into manure is known as ______
(A) Recycling (B) Biodegrading
(C) Composting (D) Trashing

34. Loss of consciousness due to land pollution is caused by______
(A) Cadmium (B) Mercury
(C) Cyanide (D) Arsenic

35. DDT Causes ______
(A) Radioactive Pollution
(B) Soil Pollution
(C) Noise Pollution
(D) Thermal Pollution

36. Amount of Lead in the soil prescribed by Environmental quality standards
(A) 0.001mg/l (B) 0.05 mg/l
(C) 0.01mg/l (D) 0.10 mg/l

37. Real estate development can cause more of ________
(A) Water pollution (B) Soil Pollution
(C) Air pollution (D) Noise Pollution

38. Which organism can be used to digest and disassemble organic pollutants, including hydrocarbons, and polychlorinated biphenyls (PCBs)?
(A) Phytoplankton (B) Mushrooms
(C) Bacteria (D) Viruses

39. Industrial, municipal sewage, treatment plants are examples of ________ pollution
(A) Diffused sources (B) Point Sources
(C) Stationary Sources (D) Line Sources

40. Waste which arise from household, institutions, markets and other activities is known as ______
(A) Industrial Waste (B) Municipal Waste
(C) Hazardous Waste (D) Solid waste

41. Organic waste is decomposed by
(A) Plants (B) Microorganisms
(C) Animals (D) Marine organisms

42. Why DDT is banned?
(A) Long persistence in the environment
(B) Highly toxic to human
(C) Kills good insects like butterflies
(D) Used against mosquito killing

43. Composting offers a method of processing and recycling of ______
(A) Garbage (B) Sewage sludge
(C) Both the above (D) None of these

44. The presence of ______ contaminants would be the strongest reason for judging municipal sewage sludge unfit for use as fertilizer?
(A) Human faces (B) Ammonia
(C) Phosphates (D) Heavy metals

45. Which of the following is a consequence of acidic deposition?
(A) It increases the likelihood of low-lying ground fogs
(B) It results in offshore eutrophication, damaging coral reefs
(C) It creates rainwater that can damage skin cells or cause cancers
(D) It changes soil chemistry, leaching out important minerals

46. Of the following, the most ecologically responsible way to dispose of sewage sludge is ______
(A) Incineration
(B) Dumping into the deep trenches of the ocean
(C) Conventional landfills
(D) Treating with heat to kill harmful bacteria, removal of toxic metals, and using as fertilizer

47. Which method is the simplest and widely used waste disposal management method of solid, non-hazardous and non-radioactive waste?
(A) Incineration (B) Sanitary land filling
(C) Coagulation
(D) Dumping into the deep trenches of the ocean

48. Function of arthropods in soil is ______
(A) Stimulate microbial activity
(B) Improving water holding capacity
(C) Provide channels root growth
(D) Bury and shred plant residue

49. Function of earthworm in soil is ______
    (A) Improving water holding capacity
    (B) Provide channels root growth
    (C) Bury and shred plant residue
    (D) All the above

50. Which of the following insecticide mostly affects the soil health?
    (A) DDT (B) Malathion
    (C) Carbaryl (D) All the above

Answers Key

Q. Ans	Q. Ans	Q. Ans	Q. Ans	Q. Ans	Q. Ans	Q. Ans
1. B	2. B	3. D	4. B	5. A	6. D	7. D
8. D	9. B	10. A	11. C	12. C	13. B	14. C
15. C	16. D	17. A	18. B	19. A	20. C	21. B
22. A	23. C	24. A	25. A	26. B	27. A	28. A
29. C	30. D	31. B	32. C	33. B	34. C	35. B
36. C	37. B	38. B	39. B	40. B	41. B	42. A
43. C	44. D	45. D	46. D	47. B	48. A	49. D
50. D						

◈

CHAPTER 5

HEAVY METAL POLLUTION

1. Heavy metals, counted as; Mercury, Lead, Cadmium, Chromium and ___________
   (A) Carbon (B) Arsenic
   (C) Oxygen (D) Helium

2. Cadmium causes endocrine disruption in _______
   (A) Fishes (B) Seals
   (C) Whales (D) Silver sea breams

3. PCBs' are widely used in _________
   (A) Aero planes (B) Building materials
   (C) Electrical equipments (D) Industries

4. In sea, dissolved mercury ions are easily absorbed in ________
   (A) Plants (B) Sediments
   (C) Animals (D) Fats

5. Which of the following is FALSE about mercury used in gold mining?
   (A) Elemental (metallic) mercury is used in the mining process to separate gold from sediments and ore
   (B) Vaporized mercury comes from heating the mercury-gold amalgam in gold shops
   (C) Mercury is converted into methylmercury by small fish
   (D) Methylmercury concentrates up the food chain

6. The toxicity of which of the following heavy metals leads to Liver cirrhosis?
   (A) Copper (B) Lead
   (C) Mercury (D) Zinc

7. Heavy metals at their standard state have specific gravity (density) of more than _______
   (A) $5g/cm^3$ (B) $10g/cm^3$
   (C) $15g/cm^3$ (D) $20g/cm^3$

8. Pollutants like heavy metals are ________ pollutants – i.e. they aren't broken down by bacteria etc. and are effectively permanent.
   (A) Conservative (B) Persistence
   (C) Accumulative (D) Magnification

9. Those animals feeding on bioaccumulators gain even higher inputs of contaminants and bioaccumulate even greater concentrations and so on; with animals at the highest trophic level obtaining highest concentrations is known as ________
   (A) Biomagnification (B) Bioaccumulation
   (C) Conservative (D) None of the above

10. Cadmium burns in air with a bright light, forming the oxide of ________
   (A) CdO (B) CdO_2
   (C) Both (D) None of the above

11. Canada Acute environmental problem occurred in Jintsu River Valley region of Japan - *ouch-ouch* or *itai-itai* is by ________
   (A) Cadmium (B) Mercury
   (C) Lead (D) Zinc

12. Cadmium is acutely toxic; lethal dose of Cd is ________
   (A) 1g (B) 2g
   (C) 0.1g (D) 3g

13. Chromium is from ________ word chroma refers to its many colorful compounds. It is a silvery, lustrous, brittle hard metal. It does not tarnish in air.
   (A) Greek (B) Latin
   (C) French (D) Spanish

14. Common oxidation states of chromium ________
   (A) Cr(VI) (B) Cr(III)
   (C) Both A and C (D) None

15. Carcinogenic form of chromium is ________
   (A) Cr(VI) (B) Cr(III)
   (C) Both A and C (D) None

16. People are exposed to chromium by:
   (A) Breathing (inhalation) and Eating of contaminated food)
   (B) Drinking of contaminated water
   (C) Skin contact (with chromium compounds/salts)
   (D) All of these

17. CCA is ________

(A) Chromated Copper Arsenate
(B) Cadmium Copper Arsenate
(C) Chromic Cupric Ammonium
(D) Chromate Chlorine Arsenate

18. **A waterborne mixture of metal oxides with which wood is treated using vacuum pressure impregnation process is known as ________**

(A) CCA (B) ACC
(C) Both (D) None

19. **Common minerals of arsenic is ________**

(A) Arsenopyrite (feass) (B) Ironpyrite
(C) Realgar (D) Kalgoorlieite

20. **Which form of arsenic is more toxic?**

(A) As(V) (B) As(III)
(C) As(II) (D) As(VII)

21. **Removal of arsenic from water to flow the water over activate**

(A) Alumina (Aluminum oxide)
(B) Iron oxide
(C) Charcoal powder
(D) None

22. **Lead in automobile batteries as additives in gasolines:**

(A) Tetraethyl lead $Pb(C_2H_5)_4$
(B) Tetramethyl lead $Pb(CH_3)_4$
(C) Both A and B
(D) None

23. **Lead replaces ________ in human body**

(A) Calcium (B) Iron
(C) Hydrogen (D) Oxygen

24. **The symbol mercury (Hg) from the Latin word, hydrargyrum, means**

(A) Liquid silver (B) Shinning silver
(C) White silver (D) Solid silver

25. **Toxic form of mercury is**

(A) Methylmercury (B) Ethylmercury
(C) Monomethylmercury (D) Both A and C

26. Half-life of methylmercury in human body is ______
    (A) 70 days (B) 10 days
    (C) 80 days (D) 140 days

27. Minamata disease occurred in a village in ______
    (A) Japan (B) China
    (C) USA (D) Africa

28. Minamata disease Onset symptoms were first seen in ______
    (A) Cats (B) Fish
    (C) Dog (D) Cattle

29. Sources of methylmercury is ______
    (A) Fungicide in agriculture and industry
    (B) Leaching from rocks and soil into water systems by natural processes
    (C) Preservative Thimerosa
    (D) All the above

30. Heavy metals cannot be degraded or destroyed, because heavy metals are ______ in all parts of the environment.
    (A) Persistence (B) Conservative
    (C) Accumulative (D) Magnification

31. In 1952, the first incidents of ______ poisoning appear in the population of Minimata Bay in Japan,
    (A) Mercury (B) Cadmium
    (C) Lead (D) Chromium

32. Minamata disease caused by consumption of ______ contaminated with mercury
    (A) Fish (B) Cattle
    (C) Pork (D) Chicken

33. Find the properties of heavy metal?
    (A) Have high densities (B) Toxic in nature
    (C) Nondegradable (D) All of the above

34. Medicinal treatment for heavy metal poisoning is done by ______
    (A) Chelates (B) Nonmetals
    (C) Drugs (D) Metal oxides

35. ______ lower IQ levels in children
    (A) Mercury (B) Cadmium
    (C) Lead (D) Chromium

36. Mercury is most toxic in ________ form
    (A) Vapour (B) Solid
    (C) Liquid (D) Fluid

37. The effective method for removal of mercury vapors is by the use of ________
    (A) Packed bed of absorbents (B) Alumina(aluminum oxide)
    (C) Iron oxide (D) Charcoal powder

38. Zinc in water (from wastewater of industrial plants) increases the ________ of waters.
    (A) Acidity (B) Alkalinity
    (C) Salinity (D) Sodicity

39. Soils that are acidified enhance the ________ uptake by plants
    (A) Mercury (B) Cadmium
    (C) Lead (D) Chromium

40. Sewage containing mercury is released by ________ works into Minimata Bay in Japan.
    (A) Chisso's chemicals (B) Basel chemicals
    (C) Sandoz chemicals (D) All of them

41. "Pollution of the Rhine at Basel / Sandoz" killed fish over a strech of 100 km by ________
    (A) Mercury (B) Cadmium
    (C) Lead (D) Chromium

42. Heavy metals are associated with myriad adverse health effects, allergic reactions
    (A) Beryllium, Chromium (B) Lead
    (C) Arsenic (D) Hexavalent chromium

43. Neurotoxicity affects nervous system by heavy metals like ________
    (A) Beryllium, Chromium (B) Lead
    (C) Arsenic (D) Hexavalent chromium

44. Nephrotoxicity affects kidneys by heavy metals like ________
    (A) Mercuric chloride (B) Cadmium chloride
    (C) Both A and B (D) Arsenic, hexavalent chromium

45. Lead, cadmium and arsenic are more toxic in their ________ form
    (A) Cationic (B) Vapour
    (C) Anionic (D) Liquid

46. ________ is/are the main source of environmental cadmium
    (A) Coal burning (B) Incineration
    (C) Both A and C (D) None

47. ________ is/are major sources of arsenic and mercury indoors
    (A) Fungicides (B) Insecticides
    (C) Both A and D (D) Pesticides

48. As trace elements, which among the following heavy metals are essential to maintain the metabolism of the human body?
    (A) Copper (B) Selenium
    (C) Zinc (D) All of the above

49. ________ is the only common metal which is liquid at ordinary temperatures.
    (A) Copper (B) Selenium
    (C) Zinc (D) Mercury

50. ________ are known to be more resistant to cadmium poisoning
    (A) Saltwater organisms (B) Freshwater organisms
    (C) Terrestrial organisms (D) None

Answers Key

Q. Ans	Q. Ans	Q. Ans	Q. Ans	Q. Ans	Q. Ans	Q. Ans
1. B	2. D	3. C	4. B	5. C	6. A	7. A
8. A	9. A	10. A	11. A	12. A	13. A	14. C
15. A	16. D	17. A	18. A	19. A	20. B	21. A
22. C	23. A	24. A	25. D	26. A	27. A	28. A
29. D	30. A	31. A	32. A	33. D	34. A	35. C
36. A	37. A	38. A	39. B	40. A	41. A	42. A
43. B	44. C	45. A	46. C	47. C	48. D	49. D
50. A						

CHAPTER 6

WATER POLLUTION

1. **Water pollution can be occurred from ______**
   (A) Point source (B) Non-point source
   (C) both (D) none of the above
2. **Urbanization generally leads to higher ______ concentrations in urban catchments.**
   (A) Potassium (B) Magnesium
   (C) Phosphorus (D) Calcium
3. **Which of the following is caused by lack of water hygiene?**
   (A) Leprosy (B) Conjunctivitis
   (C) Trachoma (D) Diarrhea
4. **Which of the following is not a water borne disease?**
   (A) Typhoid (B) Scabies
   (C) Cholera (D) Hepatitis
5. **Freshwater lakes are most often limited by ______**
   (A) Nitrogen (B) Phosphorus
   (C) Carbon (D) None of the above
6. **The process of nutrient enrichment of water body is termed as ______**
   (A) Eutrophication (B) Limiting nutrients
   (C) Enrichment (D) Schistosomiasis
7. **The source of Arsenic in water is ______**
   (A) Industrial waste (B) Fertilizers
   (C) Phosphate rocks (D) All of the above
8. **Fluorides in water can be removed by ______**
   (A) Reverse osmosis
   (B) Lime softening
   (C) Ion exchange
   (D) All of the above

9. Which of the following cause mottled and discoloured teeth in children?
   (A) Fluorides (B) Chlorides
   (C) Hardness (D) All of the above

10. According to WHO, the soft water has 0 to ________ milligram per liter of hardness as $CaCO_3$.
    (A) 30 (B) 60 (C) 90 (D) 120

11. Permanent hardness in water is caused by ________
    (A) Magnesium carbonate (B) Magnesium bicarbonate
    (C) Magnesium sulphate (D) All of the above

12. Temporary hardness is caused due to ________
    (A) Magnesium carbonate (B) Calcium sulphate
    (C) Magnesium sulphate (D) Magnesium chloride

13. The following cause alkalinity as well hardness in natural water.
    (A) Calcium carbonate (B) Calcium bicarbonate
    (C) Magnesium carbonate (D) All of the above

14. The following cause alkalinity in natural water.
    (A) Potassium carbonate (B) Potassium bicarbonate
    (C) Sodium carbonate (D) All of the above

15. According to the United States Geological Survey, water having less than 1000 mg/litre of total dissolved solids is
    (A) Fresh water (B) Slightly saline
    (C) Moderately saline (D) Brine water

16. The following unit is not used to measure turbidity of water?
    (A) NTU (B) ATU (C) JTU (D) FTU

17. The water temperature should preferably be less than ________ degree Celsius.
    (A) 10 (B) 15 (C) 25 (D) 30

18. A technique used to determine the concentration of odour compounds in a sample is known as ________
    (A) Stripping (B) Settling
    (C) Flushing (D) Chlorination

19. In filtration, the amount of dissolved solids passing through the filters is ________
    (A) Difference between total solids and suspended solids
    (B) Sum of total solids and suspended solids

(C) Independent of suspended solids
(D) None of the above

20. The Total dissolved solids (TDS) can be reduced by the following method ________
(A) Distillation (B) Reverse osmosis
(C) Ion exchange (D) All of the above

21. ________ is highly dangerous for human health causing respiratory cancer, arsenic skin lesion from contaminated drinking water.
(A) Arsenic (B) Cadmium
(C) Lead (D) Fluorine

22. ________ reaches human body through food crop from soil irrigated by affected effluents.
(A) Arsenic (B) Cadmium
(C) Lead (D) Fluorine

23. The critical pH for most of the aquatic species is ________
(A) 5.0 (B) 4.0 (C) 6.0 (D) 8.0

24. The B.O.D of unpolluted fresh water is usually below ________
(A) 1mg/l (B) 400 mg/l
(C) 10 mg/l (D) 0.5mg/l

25. The permitted B.O.D of treated effluent ________
(A) 1mg/l (B) 30 mg/l
(C) 10 mg/l (D) 500 mg/l

26. The nutrient enrichment is due to the activities of humans, it is referred to as ________
(A) Cultural eutrophication (B) Eutrophication
(C) Both A and B (D) None of the above

27. With rapid industrialization and urbanization, the water requirement for energy and industrial use is estimated to rise to about ________ of the total requirements in 2025
(A) 10 per cent (B) 18 per cent
(C) 20 per cent (D) 50 per cent

28. Examples of organic pollutants ________
(A) DDT
(B) Lindane
(C) Polychlorinated biphenyls (PCBs)
(D) All of the above

29. Examples of inorganic pollutants ______
(A) Arsenic
(B) Mercury
(C) Copper
(D) All of the above

30. Water stored in underground aquifers is known as ______
(A) Surface water
(B) Groundwater
(C) Bore water
(D) None of the above

31. If pollution comes from a single location, such as a discharge pipe attached to a factory, it is known as ______
(A) Point-source pollution.
(B) Non point source pollution
(C) Single source pollution
(D) None of the above

32. A great deal of water pollution happens not from one single source but from many different scattered sources. This is called ______.
(A) Point-source pollution.
(B) Non point source pollution
(C) Single source pollution
(D) None of the above

33. There are ______ main ways of measuring the quality of water.
(A) One
(B) Two
(C) Five
(D) Three

34. If the chemicals are dangerous or the concentrations are too great in water source like pond, river then it is ______ of water quality.
(A) Chemical indicators
(B) Biological indicators
(C) Physical indicators
(D) None of the above

35. If the river supports no fish life at all, then it is ______ of water quality.
(A) Chemical indicators
(B) Biological indicators
(C) Physical indicators
(D) None of the above

36. Sewage and fertilizers can cause a massive increase in the growth of algae or plankton that overwhelms huge areas of oceans, lakes, or rivers. This is known as a ______
(A) Harmful algal bloom
(B) Hab
(C) Red tide
(D) All of the above

37. Over ________ percent of oil pollution at sea comes from routine shipping and from the oil people pour down drains on land.

(A) 40 (B) 50 (C) 70 (D) 90

38. Clean Water Act ________

(A) 1972 (B) 1974 (C) 1988 (D) 1971

39. Safe Drinking Water Act ________

(A) 1972 (B) 1974 (C) 1988 (D) 1971

40. Mercury and cadmium are still used in ________

(A) Pipes (B) Batteries
(C) Plastics (D) Fuel

41. Theme for the 2019 world water day is ________

(A) No Water. No Life. No Blue. No Green
(B) Water is The Driving Force of all Nature.
(C) Thousands Have Lived Without Love, Not One Without Water
(D) Leaving no one behind

42. World water day on ________

(A) March 22 (B) May 2
(C) January 20 (D) March 21

43. ________ is the result of contaminated water from one country spilling into the waters of another

(A) Transboundary pollution
(B) Oil spill pollution
(C) Industrial pollution
(D) Air pollution

44. ________ pollution and its debris consumption majorly lead to the deaths of millions of marine species like sea turtles, sea birds, fishes etc.

(A) Oil spill (B) Industrial pollution
(C) Plastic (D) Water pollution

45. Expand CPCB?

(A) Central pollution control board
(B) Central pollutants control board
(C) Contaminant plastics control bench
(D) None of the above

46. Headquarters of CPCB?
    (A) Lucknow (B) Bombay
    (C) Delhi (D) Chennai

47. In which year the NGT act came into force?
    (A) National Green Tribunal Act 2008
    (B) National Green Tribunal Act 2010
    (C) National Green Tribunal Act 2002
    (D) National Green Tribunal Act 2011

48. NGT 2010 draws inspiration from the India's constitutional provision of________ article, which assures the citizens of India the right to a healthy environment.
    (A) 14 (B) 21
    (C) 34 (D) 10

The Tamil Nadu Prevention and Control of Water Pollution Board was constituted by the Government of Tamil Nadu during_______
    (A) 1982 (B) 1970
    (C) 1980 (D) 1990

50. Which is the most polluted river in India?
    (A) Chenab river (B) Yamuna river
    (C) Ganges river (D) Jhelum river

Answers Key

Q.	Ans	Q.	Ans	Q.	Ans	Q.	Ans	Q.	Ans	Q.	Ans	Q.	Ans
1.	C	2.	C	3.	D	4.	B	5.	B	6.	A	7.	D
8.	D	9.	A	10.	B	11.	C	12.	A	13.	D	14.	D
15.	A	16.	B	17.	C	18.	A	19.	A	20.	D	21.	A
22.	B	23.	C	24.	A	25.	B	26.	A	27.	B	28.	D
29.	D	30.	B	31.	A	32.	B	33.	B	34.	A	35.	B
36.	D	37.	C	38.	A	39.	B	40.	B	41.	D	42.	A
43.	A	44.	C	45.	A	46.	C	47.	B	48.	B	49.	A
50.	C												

◈

CHAPTER 7

WATER POLLUTION AND CONTROL

1. **Water bodies with extreme eutrophication will not help in the growth of ________**

   (A) Algae (B) Fishes
   (C) Pollution (D) Nutrients

2. **In Sindh, Pakistan which lake destroyed animals and birds due to high chemicals and salinity release?**

   (A) Vembanad lake (B) Manchar lake
   (C) Nathsagar lake (D) Banjosa lake

3. **The ground water of Bangaldesh is poisoned by ________**

   (A) Arsenic (B) Antimony
   (C) Plladium (D) Bismuth

4. **Fish and bioaccumulation ________.**

   (A) Mercury (B) Bismuth
   (C) Palladium (D) Chlorine

5. **Which of the following is a biological contaminant in water?**

   (A) *E. Coli* (B) Chromium
   (C) Arsenic (D) Industries

6. **Presence of bacteria and microorganisms in water will cause ________ problems in human and animals.**

   (A) Indigestion (B) Intestinal tract
   (C) Brain tumour (D) Cancer

7. **The causing organism for hepatitis is ________**

   (A) Bacteria (B) Virus
   (C) Protozoa (D) Helminth

8. **________ causes amoebic diseases.**

   (A) Virus (B) Bacteria
   (C) Helminth (D) Protozoa

9. Bacteria in water causes ___________.
   (A) Malaria (B) Typhoid
   (C) Dengue (D) Chicken guinea

10. Helminth in water causes ___________
   (A) Hook worm (B) Amoebic dysentery
   (C) Cholera (D) Typhoid

11. Which of the following is an important requirement for the aquatic life?
   (A) Dissolved nitrogen (B) Dissolved chlorine
   (C) Dissolved oxygen (D) Dissolved methane

12. Disappearance of plants and animals from water body is due to the depletion of ________ in water.
   (A) Nitrogen (B) Chlorine
   (C) Oxygen (D) Ozone

13. How many types of organic matter present in water?
   (A) Two (B) Three
   (C) Four (D) Five

14. Domestic sewage may have BOD value is about
   (A) 1600 (B) 16000
   (C) 160 (D) 160000

15. The following is the most practiced domestic water treatment
   (A) Aerobic
   (B) Anaerobic
   (C) Both Aerobic and Anaerobic
   (D) E. Coli

16. The BOD value of treated industrial effluent should be below ________ for safe discharge
   (A) 10 – 100 mg L^{-1} (B) 200 mg L^{-1}
   (C) 30 mg L^{-1} (D) 40 mg L^{-1}

17. The COD value of treated industrial effluent should be below ________ for safe discharge
   (A) 750 mg L^{-1} (B) 500 mg L^{-1}
   (C) 300 mg L^{-1} (D) 250 mg L^{-1}

18. In domestic water treatment, after the primary treatment ___________ is done.
   (A) Screening (B) Sedimentation
   (C) Aerobic process (D) Anaerobic process

19. In the domestic water treatment process, when air is sent through the active sludge, ________ released.
(A) Oxygen (B) Carbondioxide
(C) Nitrogen (D) Chlorine

20. Anaerobic treatment process ________
(A) Sedimentation (B) Screening
(C) Catalysis (D) Fermentation

21. In final step of the domestic water treatment process, the effluent contains ________ BOD.
(A) 100 ppm (B) 150 ppm
(C) 200 ppm (D) 25 ppm

22. Which among the following is an AEROBIC process?
(A) Activated sludge process
(B) Sludge thickening process
(C) Sedimentation
(D) Screening

23. Which of the following act enacted after the Exxon Valdez Oil Spill?
(A) Oil Pollution Act, 1990
(B) The Environment (Protection) Act, 1986
(C) Air (Prevention and Control of Pollution Act, 1981
(D) The Indian Forest Act, 1927

24. In which year Woburn, Massachusetts Contamination occurred?
(A) 1970 (B) 1976
(C) 1980 (D) 1990

25. What spilled in the Woburn, Massachusetts Contamination causing childhood leukaemia?
(A) Methyl Iso Cyanate
(B) Trichloroethylene and perchloroethylene
(C) Mercury
(D) Lead

26. Which of the following component caused rashes, dizziness, nausea and vomiting by contaminating Elk River?
(A) Methyl Iso Cyanate
(B) Crude 4-methylcyclohexanemethanol
(C) TCE
(D) PCE

27. The largest accidental marine water petroleum oil spill?
    (A) Deep water horizon oil spill
    (B) Atlantic Empress
    (C) Nowruz Field platform
    (D) Amoco Cadiz

28. Major contributor in polluting the river Yamuna
    (A) New Delhi (B) Haryana
    (C) Uttarakhand (D) Himachal Pradesh

29. Legionnaire's disease in the city of Flint due to high blood ______ levels due to the contamination of ______ in the water.
    (A) Lead and Lead
    (B) Carbon monoxide and Lead
    (C) Nitrite and Lead
    (D) Chlorine and Lead

30. Hinkley groundwater contamination in California was due to
    (A) Chromium VI (B) Lead
    (C) Nitrite (D) Chlorine

31. What is the desirable limit of fluorine in drinking water as per the Indian standard?
    (A) 1.0 mg/L (B) 2.0 mg/L
    (C) 3.0 mg/L (D) 4.0 mg/L

32. What may happen if the fluorine limit is more than the IS drinking water limit?
    (A) Fluorosis (B) Mathaemogolobin
    (C) Toxic (D) Carcinogenic

33. What is the desirable limit of nitrate as per the Indian standard drinking water specification?
    (A) 45 mg/L (B) 20 mg/L
    (C) 30 mg/L (D) 50 mg/L

34. What may happen if the nitrate limit is more than the IS drinking water limit?
    (A) Fluorosis (B) Mathaemogolobinemea
    (C) Toxic (D) Carcinogenic

35. What is the desirable limit of Mercury as per the Indian standard drinking water specification?
    (A) 0.001mg/L (B) 0.002 mg/L
    (C) 0.003 mg/L (D) 0.004 mg/L

36. What is the desirable limit of Cadmium as per the Indian standard drinking water specification?
    (A) 0.01mg/L (B) 0.02 mg/L
    (C) 0.03 mg/L (D) 0.04 mg/L
37. What is the desirable limit of Lead as per the Indian standard drinking water specification?
    (A) 0.05mg/L (B) 0.02 mg/L
    (C) 0.03 mg/L (D) 0.04 mg/L
38. What is the desirable limit of Arsenic as per the Indian standard drinking water specification?
    (A) 0.05mg/L (B) 0.02 mg/L
    (C) 0.03 mg/L (D) 0.04 mg/L
39. What is the desirable limit of dissolved solid as per the Indian standard drinking water specification?
    (A) 500 mg/L (B) 250 mg/L
    (C) 150 mg/L (D) 50 mg/L
40. What may happen if the aluminium limit is more than the IS drinking water limit?
    (A) Dementia (B) Memory loss
    (C) Both A and B (D) None of the above
41. Which of the following is also known as gastroentertis, causing diarrhea, vomiting, fever and stomach pain by viral or bacterial infection through water?
    (A) Stomach Flu (B) Memory loss
    (C) Blue baby syndrome (D) None of the above
42. Which type of hepatitis is caused as a result of ingestion of contaminated water and food?
    (A) Hepatitis A (B) Hepatitis D
    (C) Hepatitis A and E (D) None of the above
43. The disease caused by the *Giardia lamblia* a microscopic pathogen infected to small intestine through contaminated food and water?
    (A) Hepatitis (B) Giardiasis
    (C) Stomach Flu (D) None of the above
44. Disease caused by eating or drinking feacally contaminated food or water is ______
    (A) Silicosis (B) Giardiasis
    (C) Amoebiasis (D) None of the above

45. ________ is resistant to disinfection and affects immune system and it is the cause of diarrhoea and vomiting in humans
(A) Poliomyelitis (B) Cryptosporidium
(C) Amoebiasis (D) None of the above

46. According to WHO, diarrheal cases are about 4 billion and results in 2.2 million deaths Giardiasis is caused by ________
(A) Encephalitis (B) Gastroenteritis
(C) *Giardia lamblia* (D) None of the above

47. Gastroenteritis is caused by different viruses including ________
(A) Rotaviruses (B) Adenoviruses
(C) Calciviruses and Norwalk virus (D) All the above

48. Health risk associated with polluted water includes different diseases such as ________
(A) Respiratory disease
(B) Cancer and Diarrheal disease
(C) Neurological disorder and Cardiovascular disease
(D) All the above

49. Nitrogenous chemicals are responsible for ________
(A) Blue baby syndrome (B) Minamata disease
(C) Respiratory disease (D) None of the above

50. ________ was in the waste product dumped into Minamata Bay on a massive scale by a chemical plant which results in very high fever, convulsions, psychosis, loss of consciousness, coma, and finally death to human being.
(A) Mercury (B) Lead
(C) Arsenic (D) All the above

Answers Key

Q. Ans	Q. Ans	Q. Ans	Q. Ans	Q. Ans	Q. Ans	Q. Ans
1. B	2. B	3. A	4. A	5. A	6. B	7. B
8. D	9. B	10. A	11. C	12. C	13. A	14. C
15. A	16. C	17. D	18. A	19. B	20. D	21. D
22. A	23. A	24. A	25. B	26. B	27. A	28. A
29. A	30. A	31. A	32. A	33. A	34. B	35. A
36. A	37. A	38. A	39. A	40. A	41. A	42. C
43. B	44. C	45. B	46. C	47. D	48. D	49. A
50. A						

◆

CHAPTER 8

THERMAL AND RADIOACTIVE POLLUTION

1. **The excess of heat dissipated into air or water from the industries increases the temperature of the system and hence, causes ________.**
   (A) chemical pollution (B) air pollution
   (C) water pollution (D) thermal pollution

2. **The effect of thermal pollution is more prominently marked in ________**
   (A) aquatic system (B) terrestrial system
   (C) grassland (D) forest

3. **The industries like iron and steel plants, petroleum refineries, nuclear reactors, electric power plants etc. use large amount of water for ________**
   (A) heating process (B) cooling purposes
   (C) bleaching (D) none of the above

4. **When an increase in temperature of the aquatic body affects and disrupts the normal activities of the aquatic living organisms, the process is known as ________**
   (A) chemical pollution (B) air pollution
   (C) water pollution (D) thermal pollution

5. **Increase in temperature of the aquatic bodies by ________ degrees Celsius becomes injurious to the aquatic life.**
   (A) 8 to 10 (B) 1 to 2
   (C) 50 to 55 (D) 30 to 35

6. **Some important sources of thermal pollution ________**
   (A) Nuclear reactors
   (B) Industrial wastes
   (C) Hydro-electric power plant / Thermal power plant
   (D) All the above

7. Effect of thermal pollution;
   (A) It reduces the dissolved oxygen content of water.
   (B) It influences reproductive cycle, digestion rate, respiration rate and many enzymatic activities of living organisms.
   (C) It favours the growth of certain bacteria and pathogens.
   (D) All the above

8. Geothermal vents and hot springs introduce excess ______ into bodies of water.
   (A) Cold (B) Pollution
   (C) Heat (D) Soil

9. Soil erosion, deforestation, and ______ from paved areas are other sources of hot water.
   (A) Runoff (B) Afforestation
   (C) Litter (D) None of the above

10. Water on hot paved surfaces gets hot, then runs off into nearby bodies of water, and thereby raising the water ______.
   (A) Pressure (B) Temperature
   (C) Pollution (D) Density

11. ______ can also be a source of thermal shock because the relatively small and shallow bodies of water can absorb quite a bit of heat energy from the sun.
   (A) Retention ponds (B) River
   (C) Lakes (D) sea

12. Warm water holds ______ than cool water.
   (A) More oxygen (B) More CO_2
   (C) Less oxygen (D) less CO_2

13. The decreased oxygen can cause ______ that pose a threat to aquatic plants and animals.
   (A) Death of algae blooms (B) nutrient enrichment
   (C) increase in food (D) none of the above

14. Thermal pollution causes ______ which contributes to a dramatic loss of biodiversity.
   (A) Growth in population (B) Migration
   (C) Decrease in temperature (D) None of the above

15. ______ is an almost inevitable side effect of using water for cooling.
   (A) Chemical pollution (B) Air pollution
   (C) noise pollution (D) land pollution

16. Nuclear power plants release _______ to water.

(A) Acidic substance (B) Alkaline substance
(C) Radioactive substance (D) Neutral pH

17. _______ has also been observed when a power plant or factory waste is dumped into coastal water.

(A) Coral reef bleaching (B) Increase of minerals
(C) CO_2 emission (D) None of the above

18. Molluscs, sponges, and crustaceans which die at temperatures more than ___degrees Celsius.

(A) 2 (B) 10 (C) 37 (D) 74

19. In the case of _______ from reservoirs into warmer receiving canals, streams, rivers, or ponds, there can be substantial impacts on fish, mostly in the egg and larvae stage.

(A) cold water injection (B) salt intrusion
(C) recycled water (D) chlorinated water

20. A large increase in temperature can lead to the denaturing of life-supporting enzymes by breaking down _______ within the quaternary structure of the enzymes.

(A) Covalent bond
(B) Ionic bond
(C) hydrogen- and disulphide bonds
(D) aromatic

21. Abrupt change in water temperature, either an increase or decrease, known as _______.

(A) thermal shock (B) water heat
(C) water stress (D) heat stress

22. Any of two or more species of atoms or nuclei that have same number of neutrons are called _______

(A) Isotopes (B) Isobars
(C) Isotones (D) All of the above

23. _______ man-made bodies of water designed for cooling by evaporation, convection, and radiation.

(A) cooling ponds (B) percolation pond
(C) artificial lakes (D) radiative pond

24. _______ transfer waste heat to the atmosphere through evaporation and/or heat transfer.

(A) cooling towers (B) cooling ponds
(C) percolation pond (D) artificial lakes

25. ________ a process where waste heat is recycled for domestic and/or industrial heating purposes
(A) cooling towers (B) cooling ponds
(C) percolation pond (D) cogeneration

26. ________ is defined as the physical pollution of living organisms and their environment as a result of release of radioactive substances into the environment during nuclear explosions.
(A) radioactive pollution (B) thermal pollution
(C) heat pollution (D) noise pollution

27. The proportion of radioactive pollution is ______ of the total energy of the explosion.
(A) 10% (B) 15%
(C) 50% (D) 55%

28. ________ is the form of radiation that has a short wavelength and a high frequency.
(A) Ionizing radiation (B) Non ionizing radiation
(C) Frequency (D) None of the above

29. ________ radioactive pollution is the type of pollution that occurs during nuclear tests or during experimental tests on radioactive substances.
(A) Occasional (B) Continuous
(C) Accidental (D) None of the above

30. ________ radioactive pollution is the type of pollution constantly coming from uranium mines, nuclear reactors, and test laboratories, where the radioactive contaminants are always present
(A) Occasional (B) Continuous
(C) Accidental (D) None of the above

31. ________ radioactive pollution is the type of pollution that occurs when certain experiments involving dangerous substances fail, and the substances used for experimentation get out of control.
(A) Occasional (B) Continuous
(C) Accidental (D) None of the above

32. ________ is used as fuel for nuclear reactors.
(A) Radon (B) Caesium 137
(C) Uranium 235 (D) Strontium 90

33. ________ is used for radiation therapy in medicine (to treat cancer).
(A) Radon (B) Caesium 137
(C) Uranium 235 (D) Strontium 90

34. ________ is used for thermoelectric generators and portable power sources for space vehicles, weather stations etc.
(A) Radon (B) Caesium 137
(C) Uranium 235 (D) Strontium 90

35. ________ is used as traceable radiation that can indicate various medical parameters.
(A) Radon (B) Caesium 137
(C) Uranium 235 (D) Chromium 51

36. ________ is used for lightning rods.
(A) Radon (B) Caesium 137
(C) Uranium 235 (D) Radium 225

37. ________ represents about 55% of the natural radiation.
(A) Radon (B) Caesium 137
(C) Uranium 235 (D) Strontium 90

38. The worst nuclear reactor accident occurred in ________ in Chernobyl, Ukraine.
(A) 1976 (B) 1986
(C) 1886 (D) 1896

39. ________ is a unit representing the rate of radioactive decay.
(A) Curie (B) Radon
(C) Rem (D) gamma

40. Miners are exposed to relatively high doses of radon-222 and consequently have relatively high rates of ________.
(A) Skin cancer (B) Throat cancer
(C) lung cancer (D) none of the above

41. Radioactivity from the fallout of nuclear weapons testing and from nuclear power plants make up less than ________% of the total radiation dose.
(A) 1 (B) 10 (C) 50 (D) 0.5

42. ________ is damage to our body caused by a large dose of radiation often received over a short period of time (acute).
(A) Radiation sickness (B) acute radiation syndrome
(C) radiation poisoning (D) All of the above

43. ________ are atoms of an element that have the same atomic number but different in atomic mass.
(A) Isotopes (B) Isobars
(C) Isotones (D) All of the above

44. The half-life period of Uranium-235 is ________
(A) 700 million years (B) 5570 years
(C) 3.8 days (D) 3 minutes

45. The alpha particles can travel only up to ________ cm in air and are least penetrating.
(A) 10 to 50 (B) 5 to 8
(C) 50 to 80 (D) 1 to 2

46. The important man-made sources of radioactive pollution are
(A) Radioactive wastes generated during the mining and processing of radioactive minerals.
(B) Nuclear fallout due to the testing of nuclear weapons.
(C) Medical use such as X-rays, radiotracers for detecting and curing cancers, etc.
(D) All of the above

47. BARC stands for
(A) Bhabha Atomic Research Centre
(B) Bhabha atomic radiactive centre
(C) Bhabha atomic radiactive control
(D) Bhabha atomic radiation centre

48. A line on a weather map that joins place that have the same air pressure is ________
(A) Isotopes (B) Isobars
(C) Isotones (D) All of the above

49. Fukushima nuclear disaster in Japan happened in ________
(A) 2010 (B) 2011
(C) 2009 (D) 2014

50. The international radiation symbol also known as ________
(A) Delta (B) Gamma
(C) Trefoil (D) radon

Answers Key

Q.	Ans	Q.	Ans	Q.	Ans	Q.	Ans	Q.	Ans	Q.	Ans	Q.	Ans
1.	D	2.	A	3.	B	4.	D	5.	A	6.	D	7.	D
8.	C	9.	A	10.	B	11.	A	12.	C	13.	A	14.	B
15.	A	16.	C	17.	A	18.	C	19.	A	20.	C	21.	A
22.	C	23.	A	24.	A	25.	D	26.	A	27.	B	28.	A
29.	A	30.	B	31.	C	32.	C	33.	B	34.	D	35.	D
36.	D	37.	A	38.	B	39.	A	40.	C	41.	D	42.	D
43.	A	44.	A	45.	B	46.	D	47.	A	48.	B	49.	B
50.	C												

CHAPTER 9

NOISE POLLUTION

1. ________ is the propagation of noise with harmful impact on the activity of human or animal life.
   (A) Noise pollution
   (B) Air pollution
   (C) Chemical pollution
   (D) Environmental pollution

2. Noise is measured in ________
   (A) Dobson
   (B) Decibel
   (C) Ds'
   (D) mm

3. High noise levels can contribute to ______________ in humans and an increased incidence of coronary artery disease
   (A) Brain effect
   (B) Lung effect
   (C) Cardiovascular effect
   (D) Kidney effect

4. Noise pollution can cause ________
   (A) hypertension and high stress levels
   (B) tinnitus and hearing loss
   (C) sleep disturbances
   (D) all the above

5. According to a 2019 review of the existing literature, noise pollution was associated with ______________
   (A) Faster cognitive
   (B) Slower cognitive decline
   (C) Faster cognitive decline
   (D) Option b and c

6. Sound becomes unwanted when it ___________
   (A) interferes with normal activities such as sleep or conversation
   (B) disrupts or diminishes one's quality of life
   (C) both A and B
   (D) none of the above

7. Noise-induced hearing loss can be caused by prolonged exposure to noise levels above ________ decibels

(A) 85 (B) 75
(C) 95 (D) 65

8. ________ is noise that is produced continuously by machinery that keeps running without interruption. This could come from factory equipment, heating or ventilation systems

(A) Continuous noise (B) Intermittent noise
(C) High frequency noise (D) Low frequency noise

9. ________ is a noise level that increases and decreases rapidly. This might be a freight train passing by, factory equipment that operates in cycles or aircraft overhead.

(A) Increased noise (B) Decreased noise
(C) Rapid noise (D) Intermittent noise

10. ________ is most commonly associated with the construction and demolition industry. This sudden burst of noise can startle you by its fast and surprising nature.

(A) External noise (B) High frequency noise
(C) Impulsive noise (D) Low frequency noise

11. ________ makes up part of the fabric of our daily soundscape. This is also the hardest type of noise to reduce at source, so it can easily spread for miles around.

(A) Low frequency noise (B) High frequency noise
(C) External noise (D) Low frequency voice

12. ________ are commonly created by explosions or construction equipment such as pile drivers.

(A) Continuous noise (B) Intermittent noise
(C) Low frequency noise (D) Impulsive noise

13. Higher ambient noise levels also cause animals to vocalize more loudly, which is called ________

(A) Autistic spectrum (B) Lombard effect
(C) Animal hazard (D) Frequency effect

14. Noise pollution may have caused the death of certain species of whales that ________ themselves after being exposed to the loud sound of military sonar

(A) Beached (B) Migrate
(C) Fight (D) jumped

15. In recent years ________ programs and initiatives have arisen in an effort to combat occupational noise exposures.

(A) Earphone (B) Muffler
(C) Buy quiet (D) Option A and B

16. Noise from roadways and other urban factors can be mitigated by ________

(A) Urban Planning (B) Better design of roads
(C) Option a only (D) Option a and b

17. Roadway noise can be reduced by

(A) limitation of vehicle speeds and alteration of roadway surface texture,
(B) limitation of heavy vehicles,
(C) use of traffic controls that smooth vehicle flow to reduce braking and acceleration, and tire design
(D) all the above

18. ________ can be reduced by using quieter jet engines

(A) Air pollution (B) Sound pollution
(C) Noise pollution (D) Aircraft noise

19. Awaaz Foundation is an Indian NGO working at ________ to control noise pollution from various sources through advocacy, public interest litigation, awareness, and educational campaigns since 2003.

(A) Mumbai (B) Gujrat
(C) Chennai (D) Bangalore

20. The Supreme Court of India had banned playing of music on loud speakers after ______

(A) 9.00 pm (B) 12.00 pm
(C) 1.00 am (D) 10.00 pm

21. In 2015, The National Green Tribunal directed authorities in ________ to ensure strict adherence to guidelines on noise pollution, saying noise is more than just a nuisance as it can produce serious psychological stress.

(A) Mumbai (B) Karnataka
(C) Delhi (D) Chennai

22. The Noise Control Act was enacted in the year ________

(A) 1962 (B) 1980
(C) 1982 (D) 1972

23. In the past, ________ coordinated all federal noise control activities through its Office of Noise Abatement and Control.
(A) Environmental Protection Agency
(B) Environmental pollution act
(C) Environmental pollution rule
(D) All the above

24. Environmental noise is an accumulation of noise pollution that occurs outside. This noise can be caused by ________ activities.
(A) transport (B) industrial
(C) recreational (D) All the above

25. Noise is measured in decibels (dB) using a pattern-approved ________
(A) dB meter (B) Sound level meter
(C) Sound absorber (D) Remote sensing

26. Sound production is divided into ________ categories:
1. Mechanical noise
2. Aerodynamic noise
3. Noise from aircraft systems
(A) Option 1 and 2 (B) Option 2 and 3
(C) Option 1 and 3 (D) Option 1 2 and 3

27. ________ noise is from the airflow around the surfaces of the aircraft, especially when flying low at high speeds
(A) Mechanical noise (B) Harmful noise
(C) Aerodynamic noise (D) Flying noise

28. Aircraft engines are the major source of noise and can exceed ________ Decibels (dB) during takeoff.
(A) 140 dB (B) 100 dB
(C) 200dB (D) 50 dB

29. Elevated workplace or other noise can cause
(A) hearing impairment
(B) annoyance and sleep disturbance
(C) decreased school performance
(D) all the above

30. The ________ regulates the maximum noise level that individual civil aircraft can emit through requiring aircraft to meet certain noise certification standards.

(A) Federal Aviation Administration(FAA)
(B) Noise pollution act
(C) Environmental pollution act
(D) Environmental protection agency

31. Chronic noise exposure has been associated with
(A) sleep disturbances (B) incidence of diabetes
(C) option A (D) option A and B

32. ____________ is music that is played at a high volume, often to the point where it disturbs others and/or causes hearing damage.
(A) Devotional music (B) Marriage song
(C) Radio music (D) Loud music

33. Many jurisdictions have laws defining loud music as a criminal offense, typically a _____
(A) Very dangerous (B) Misdemeanor
(C) Neglected crime (D) Option a and c

34. NIOSH is stand for
(A) National Institute for Occupational Safety and Health
(B) National Industry for Occupational Safety and Health
(C) National Institute for Occupational Safety and Hazard
(D) National Institute for Occupational Severity of Hazard

35. ________ is unwanted sound judged to be unpleasant, loud or disruptive to hearing
(A) Air pollution (B) DJ sound
(C) Industrial sound (D) Noise

36. Continual exposure to loud music can also lead to __________
(A) Tinnitus (B) Cancer
(C) Blindness (D) None of the above

37. Decibel is related to
(A) Alexander Graham Bell (B) Alexander Hamilton
(C) Alexander Fleming (D) Alexander von Humboldt

38. __________ is the addition of natural or artificial sound into an environment to cover-up unwanted sound by using auditory masking.
(A) Sound masking (B) White masking
(C) Pink masking (D) None of the above

39. Many of these techniques rely upon material science applications of constructing sound baffles or using ________ for interior spaces

(A) Earplugs (B) Sound-absorbing liners
(C) Mask (D) Soundproofing

40. ________ is the branch of engineering dealing with sound and vibration.

(A) Acoustic engineering (B) Acoustic
(C) Audacity (D) Audacity engineering

41. International Noise Awareness Day is a global campaign, founded in ________ by the Center for Hearing and Communication (CHC)

(A) 1990 (B) 1980
(C) 1890 (D) 1996

42. ________ is the process of calculating the level of noise emission using the metric dB(A).

(A) Noise calculation (B) Sound calculation
(C) Decibel calculation (D) None of the above

43. Noise calculation is done for the following reasons:

i. to calculate a current situation based on existing physics, mostly when a physical measurement at a location is impractical

ii. to predict the resulting immission levels based on a planned change, *e.g.* set up of a new machine

(A) option i (B) option i and ii
(C) option ii (D) None

44. To measure impulsive noise, ________ needed to calculate Peak values.

(A) Sound level meter with Third Octave Band analysis
(B) Sound level meter
(C) Personal noise dosimeter
(D) Option B and C

45. Noise pollution can be reduced by

(A) Lowering the volume
(B) Planting more trees
(C) Regular maintenance of vehicles and machines
(D) All the above

46. Normally noise which is less than ______ dB is tolerable (for shorter period).

(A) 115 (B) 200
(C) 150 (D) 180

47. The Noise Pollution and Abatement Act

(A) 1872 (B) 1972
(C) 1772 (D) 1672

48. International Noise Awareness Day is on ______

(A) March 25th (B) April 25th
(C) December 25th (D) March 22th

49. World Hearing Day is held on ______ each year to raise awareness to prevent deafness and hearing loss and promote ear and hearing care across the world.

(A) 1 March (B) 1 December
(C) 3 March (D) 3 December

50. The Anti-Noise Pollution Campaign in Mumbai by ______

(A) Noise foundation (B) Awaaz foundation
(C) Anti-pollution NGO (D) Pradushan foundation

Answers Key

Q. Ans	Q. Ans	Q. Ans	Q. Ans	Q. Ans	Q. Ans	Q. Ans
1. A	2. B	3. C	4. D	5. C	6. C	7. A
8. A	9. D	10. C	11. A	12. D	13. B	14. A
15. C	16. D	17. D	18. D	19. A	20. D	21. C
22. D	23. A	24. D	25. B	26. D	27. C	28. A
29. D	30. A	31. D	32. D	33. B	34. A	35. D
36. A	37. A	38. A	39. B	40. A	41. D	42. A
43. B	44. D	45. D	46. A	47. B	48. B	49. C
50. B						

◆

CHAPTER 10

Plastic Pollution, Agricultural Pollution, Light Pollution, Hazardous Waste And E-Waste

1. Plastics enter the marine environment primarily by ________
   (A) Being dumped or lost there
   (B) Debris carried in runoff
   (C) Washing out of landfills
   (D) None of the answers are correct

2. The plastic product used for making water bottles is ________
   (A) Polyethylene terephthalate (PET)
   (B) High – Density polyethylene (HDPE)
   (C) Polypropylene
   (D) Polystyrene

3. Plastic bags are made up of ________
   (A) Polyethylene terephthalate (PET)
   (B) High – Density polyethylene (HDPE)
   (C) Polystyrene
   (D) Low – Density polyethylene (LDPE)

4. The total amount of plastic waste ever produced has been recycled ________
   (A) 12% (B) 18% (C) 9% (D) 7%

5. The total amount of plastic waste that have been incinerated is ________
   (A) 45% (B) 12% (C) 52% (D) 36%

6. The total amount of plastic waste has accumulated in landfills, dumps or the natural environment ________
   (A) 82% (B) 91% (C) 68% (D) 79 %

7. The most common type of plastic waste found in the environment ______

(A) Cigarette butts (B) Drink bottles
(C) Food wrappers (D) Grocery bags

8. The amount of plastic waste being dumped in ocean every year is about ______

(A) 10 million tonnes (B) 12 million tones
(C) 8 million tones (D) 6 million tones

9. The amount of plastic waste carried out by Ganges, Brahmaputra, Meghna is______

(A) 52000 tons (B) 72845 tons
(C) 34000 tons (D) 65000 tons

10. The river which is contributing to larger amount of plastic waste in oceans is ______

(A) Chang Jiang (B) Indus
(C) Nile (D) Huang He

11. The amount of plastic waste carried out by Chang Jiang river is______

(A) 1,469,481 tons (B) 164,332 tons
(C) 124,249 tons (D) 84,792 tons

12. Plastic decomposes even after ______

(A) 10 Years (B) 50 Years
(C) 1000 Years (D) 100 Years

13. Which of the following is not the effect of modern agriculture ______

(A) Nitrate pollution (B) Eutrophication
(C) Biomagnification (D) Ozone depletion

14. The human activity, among the following, which causes maximum environmental pollution having regional and global impacts is ______

(A) Urbanization (B) Industrialization
(C) Agriculture (D) Mining

15. Electronic waste is the adverse effect of ______

(A) Agriculture (B) Transportation
(C) Industry (D) Mining

16. SPCB's are required to give notice in how many newspapers mentioning the date, time and place of public hearing

(A) 2 (B) 4 (C) 6 (D) 8

17. "Meeting the needs of the present without compromising the ability of future generation to meet their own need" is given by
(A) Brundtland (B) Mahatma Gandhi
(C) Maathai (D) Sunderlal Bahugana

18. Which of the following element is cause of e-waste?
(A) Lead (B) Cadmium
(C) Beryllium, or Brominates flame retardants
(D) All of the above

19. Which of the following are the main contributors of the e-waste in the world?
I. Refrigerators/freezers, washing machines, dishwashers
II. Small household appliances (toasters, coffee makers, irons, hairdryers)
III. Personal computers, telephones, mobile phones, laptops, printers, scanners, photocopiers
IV. Gas cylinder, chimneys & home appliances
(A) Only I, II, III (B) Only I & II
(C) Only I, III, IV (D) All of the above

20. How are electronic items dangerous?
(A) They degrade over time, releasing cancer-causing chemicals into the air.
(B) Lead and mercury in components can cause metabolic changes in users.
(C) They leach toxic metals in landfills and into ground water.
(D) They create electromagnetic fields that interfere with animal reproduction

21. Which toxic compound is not found in e-waste?
(A) Mercury (B) Cadmium
(C) Neon (D) Lead

22. What does e - waste stand for?
(A) Environment waste (B) Electronic waste
(C) Equipment waste (D) None of the above

23. What are the health hazards which can be caused by E-Waste?
(A) Lung cancer (B) DNA damage
(C) Brain (D) All of the above

24. Which country produces the most e-waste?
(A) India (B) China (C) USA (D) France

25. **How much lead is in a cathode ray tube (CRT), commonly called a picture tube**
    (A) 1.75 pounds
    (B) Between 2 and 3 pounds
    (C) 19 pounds
    (D) Between 4 and 7 pounds

26. **The most valuable part of a PC or TV is the ________**
    (A) Lead in the CRT
    (B) Circuit boards that contain silver and gold
    (C) Copper in the cathode yoke
    (D) None of the above

27. **Which of the following is the correct term related to E-waste?**
    (A) E-waste is a popular, informal name for electronic products nearing the end of their "useful life.
    (B) Computers, televisions, VCRs, stereos, copiers, and fax machines are common electronic products.
    (C) Both A and B
    (D) None of the above

28. **If water containing DDT (Dichloro-Diphenyl-Trichloroethane: $C_{14}H_9O_5$. is fed upon by cattle, it will**
    (A) get stored in the fatty tissues of animals
    (B) get stored in the liver
    (C) be excreted out with urine
    (D) get stored in the muscle fibers

29. **Along trophic level, in a food chain, insecticide amount ________**
    (A) decreases at each trophic level
    (B) remains unchanged
    (C) may increase or decrease depending on consumption
    (D) increases at each trophic level

30. **DDT stands for ________**
    (A) Dichloro-triphenyldichloroethane
    (B) Dichlorinotriammomium methane
    (C) Difluoro-diphenyl-tribromobutane
    (D) Dichloro-diphenyl tricholoroethane

31. **DDT (dichloro-diphenyl-trichloroethane: $C_{14}H_9Cl_5$ is a form of ________**
    (A) Insecticide
    (B) Herbicide
    (C) Pesticides
    (D) Parasiticides

32. Which among the following is considered as most effective pesticide?
(A) Carbamates (B) Organophosphates
(C) Organochlorines (D) All the above

33. Find the correct statement connected with DDT
(A) not a pollutant
(B) an antibiotic
(C) an antiseptic
(D) a non biodegradable agent

34. Find out the correct match
(A) Carbamates – Malathion
(B) Organophosphates – Carbofuran
(C) Organochlorine – Endosulphan
(D) None of these

35. Insecticides generally attack ________
(A) respiratory system (B) nervous system
(C) muscular system (D) circulatory system

36. When an industry removes water from a source and then returns the heated water to its source is known as ________
(A) Water pollution (B) Soil pollution
(C) Thermal pollution (D) Marine Pollution

37. Which one of the following can cause thermal pollution?
(A) Residential houses (B) Power plants
(C) Death of marine organisms (D) Oil spill

38. What is the effect of warmer temperature to the fishes?
(A) Increase the metabolism
(B) Decrease the metabolism
(C) Stabilize the metabolism
(D) Increase the solubility of oxygen

39. What is the disadvantage of control measures of thermal pollution?
(A) Leakage loss (B) Absorption loss
(C) Dilution (D) Evaporation loss

40. Which one of the following cause thermal pollution?
(A) Release of hot water (B) Organic manures
(C) Purified water (D) More number of trees

41. **How does soil erosion cause thermal pollution?**
    (A) By making natural water bodies to hold in its normal level
    (B) By polluting the water bodies
    (C) By avoiding sunlight to fall on the water bodies
    (D) By making natural water bodies to rise beyond their normal level

42. **What is a thermal shock?**
    (A) Sudden raises of temperature to abnormal level
    (B) Sudden cooling of temperature to abnormal level Purified water
    (C) Temperature don't change
    (D) Temperature change only due to environmental factors

43. **What is the main effect of thermal pollution to the oxygen solubility in water bodies?**
    (A) They increase the solubility of oxygen in water bodies
    (B) They maintain the solubility of oxygen in water bodies
    (C) They reduce the solubility of oxygen in water bodies
    (D) They don't cause any affect in solubility of oxygen to the water bodies

44. **Which of the following is the main reason for thermal pollution?**
    (A) Bio fuels (B) Organic farming
    (C) Eco friendly vehicles (D) Power plants

45. **How does an artificial lake help in solving thermal pollution?**
    (A) It stores hot water
    (B) Its stored cold water
    (C) It is used during the summer season where water scarcity cause
    (D) It helps to breed fishes

46. **Use of heat from hot water by industries is ________**
    (A) Cooling towers (B) Recycling
    (C) Heat pump (D) Co-generation

47. **How many types of cooling towers are there?**
    (A) One (B) Two
    (C) Three (D) Four

48. **What is the hazardous pollutant released from fax machines?**
    (A) Lithium (B) Selenium
    (C) Lead (D) Copper

49. Which of the following techniques can be implemented to reduce hazardous waste?

(A) Analysis

(B) Transportation

(C) Trans boundary movement

(D) Chemical exchange

50. When did Hazardous waste management act enacted in India?

(A) 1988 (B) 1989

(C) 1990 (D) 1991

Answers Key

Q. Ans	Q. Ans	Q. Ans	Q. Ans	Q. Ans	Q. Ans	Q. Ans
1. B	2. A	3. D	4. C	5. B	6. D	7. A
8. C	9. B	10. A	11. A	12. C	13. D	14. B
15. C	16. A	17. A	18. D	19. A	20. C	21. C
22. B	23. D	24. B	25. D	26. C	27. C	28. A
29. D	30. D	31. A	32. B	33. D	34. C	35. A
36. C	37. B	38. A	39. D	40. A	41. B	42. A
43. C	44. D	45. A	46. D	47. B	48. B	49. D
50. B						

CHAPTER 11

POLLUTION CONTROL

1. Trees can control ________
   (A) Water pollution (B) Air pollution
   (C) Noise pollution (D) All

2. The protocol which decided to completely phase out CFC is ________
   (A) Montreal protocol (B) Stockholm convention
   (C) Cartagena protocol (D) Kyoto protocol

3. Which of the following chemicals cause water pollution ________
   (A) Arsenic (B) Copper
   (C) Zinc (D) Sulphur
   Which of the following is correct?
   (A) Only A (B) B, C, D
   (C) A, B, C (D) A, C, D

4. Which of the following Articles of the Indian Constitution is/are related to the protection of the environment?
   (A) Article 48A (B) Article 51A(g)
   Which of the following is correct?
   (A) Only A (B) Only B
   (C) Both A & B (D) Neither A nor B

5. Air Quality Index (AQI) takes into account the presence of ________ pollutants in the environment.
   (A) 8 (B) 5
   (C) 9 (D) 6

6. National Air Quality Index was launched in the year ________
   (A) 17 September, 2013 (B) 17 September, 2014
   (C) 17 September, 2015 (D) 17 September, 2016

7. The Water (Prevention and Control of Pollution) cess Act was enacted in ______

(A) 1975 (B) 1976 (C) 1977 (D) 1978

8. Water (Prevention and Control of Pollution) Act was last amended in ______

(A) 2003 (B) 2007 (C) 2009 (D) 2011

9. The Environment Protection Act was last amended in ______

(A) 1993 (B) 1991 (C) 1997 (D) 1999

10. Air (Prevention and Control of Pollution) Act was last amended in ______

(A) 1987 (B) 1992 (C) 1978 (D) 1997

11. Ganga action plan was launched in the year ______

(A) 1986 (B) 1988 (C) 1985 (D) 1987

12. The Water Act contains ______ chapters.

(A) 7 (B) 5 (C) 6 (D) 8

13. The water act have ______ Sections.

(A) 64 (B) 65 (C) 66 (D) 67

14. The Air Act contains ______ chapters.

(A) 7 (B) 5 (C) 6 (D) 8

15. The Air act have ______ Sections.

(A) 44 (B) 54 (C) 64 (D) 74

16. The Air Act contains ______ chapters.

(A) 2 (B) 4 (C) 8 (D) 7

17. The Air act have ______ Sections.

(A) 25 (B) 12 (C) 26 (D) 14

18. The first of the major environmental protection act to be promulgated in India was ______

(A) Water act (B) Noise pollution rule

(C) Air act (D) Environmental act

19. Methods of controlling water pollution fall into ______ general categories

(A) 2 (B) 4 (C) 8 (D) 7

20. Method for controlling gaseous emissions is ______

(A) Adsorption (B) Scrubbers

(C) Catalytic converters (D) None

21. **Which chemical is present in scrubbers?**
(A) Magnesium oxide (B) Magnesium hydroxide
(C) Magnesium sulphite (D) Magnesium Chloride

22. **Methods of controlling air pollution fall into ______ general categories**
(A) 2 (B) 4 (C) 8 (D) 7

23. **A ______ is a device for removing particulates from stack gases.**
(A) Adsorption (B) Scrubbers
(C) Cyclone converter (D) None

24. **Safer method of dealing with solid wastes ______.**
(A) Landfills (B) Incineration
(C) Disposal (D) Composting

25. **The sources of manmade air pollution are______**
(A) Stationary (B) Mobile
(C) Both A & B (D) Neither A & B

26. **Wet collectors can be used to collect ______**
(A) Gaseous matter (B) Particulate matter
(C) Both A & B (D) Neither A nor B

27. **The ______ provides data and other information on the chemical composition and related physical characteristics of the atmosphere**
(A) Global Atmospheric Watch programme
(B) Global Environmental Monitoring System
(C) United Nations Environment Programme
(D) World Meteorological Organization

28. **______ types of parameters influence pollutant dispersion.**
(A) 2 (B) 4 (C) 8 (D) 7

29. **The solid waste management rules was enacted in ______**
(A) 2005 (B) 2010 (C) 2012 (D) 2016

30. **Manufacture, Storage and Import of Hazardous Chemical Rules enacted during ______**
(A) 1989 (B) 1998 (C) 1978 (D) 1988

31. **Central pollution control board was constituted during ______.**
(A) 1974 (B) 1975 (C) 1976 (D) 1977

32. **There are ______ types of industrial wastes.**
(A) 2 (B) 3 (C) 4 (D) 5

33. Secondary pollutants are formed by __________ reactions.
I. Thermal II. Chemical III. Photochemical
Which of the following is correct?
(A) I only (B) II & III only
(C) I & II only (D) I, II, III

34. __________ standards are the maximum levels consistent with an adequate safety margin.
(A) Primary (B) Secondary
(C) Tertiary (D) Quaternary

35. Dispersion of pollutants is dependent on __________
a. Emission conditions b. Transport
c. Turbulent mixing. d. Accumulation
Which of the following is correct?
(A) a only (B) b & c only
(C) A and B only (D) a, b, c

36. Sound becomes hazardous noise pollution at __________ decibels
(A) above 30 (B) above 80
(C) above 100 (D) above 120

37. The pollutant free energy resource is __________
(A) Coal (B) Petroleum
(C) Solar energy (D) Nuclear energy

38. Which of the following is not a way to conserve water?
(A) replace (B) reduce
(C) reuse (D) recycle

39. Trees help in reducing the pollution of our environment. It is promoted by ______
(A) Forest conservation day (B) Van Mahotsav
(C) Wildlife week (D) Plantation month

40. Plastic debris affects nearly __________ species worldwide through entanglement and ingestion, alters natural biological and chemical processes.
(A) 500 (B) 600
(C) 700 (D) 750

41. _____ pollution is emerging as a top threat to ocean ecosystems.
(A) Plastic (B) Soil
(C) Water (D) Noise

42. Displexia is caused by ________

(A) Cd (B) Ar (C) Pb (D) Hg

43. There are about ________ categories present in AQI.

(A) 5 (B) 6 (C) 8 (D) 10

44. Which one of the following is the transboundary problem?

(A) Water pollution (B) Air pollution
(C) Noise pollution (D) Soil pollution

45. Central Pollution Control Board is a ________ body / organization.

(A) Statutory (B) Non statutory
(C) Constitutional (D) Non Constitutional

46. Maximum permissible decibel limit for industries is ________

(A) 110 (B) 100 (C) 80 (D) 90

47. ECOMARK scheme was launched in the year ________

(A) 1991 (B) 1981 (C) 1971 (D) 1988

48. National Ambient Air Quality Standards are prescribed for a period of ________

(A) 12 hrs (B) 24 hrs (C) 36 hrs (D) 48 hrs

49. Which organization tackles marine pollution issues locally?

(A) UN (B) UNEP
(C) UNESCO (D) WMO

50. Air (Prevention and Control of Pollution) Act was passed in the year ________

(A) 1981 (B) 1977 (C) 1986 (D) 1992

Answers Key

Q. Ans	Q. Ans	Q. Ans	Q. Ans	Q. Ans	Q. Ans	Q. Ans
1. D	2. A	3. C	4. C	5. A	6. B	7. C
8. A	9. B	10. A	11. A	12. D	13. A	14. C
15. B	16. B	17. C	18. A	19. B	20. A	21. A
22. A	23. C	24. D	25. C	26. C	27. A	28. A
29. D	30. A	31. A	32. A	33. D	34. A	35. D
36. D	37. C	38. A	39. B	40. C	41. A	42. A
43. A	44. B	45. A	46. A	47. A	48. B	49. B
50. A						

◈

CHAPTER 12

SOLID WASTE MANAGEMENT

1. _________ is defined as "any garbage, refuse, sludge from a waste treatment plant, water supply treatment plant, or air pollution control facility and other discarded material, including solid, liquid, semi-solid or contained gaseous material resulting from industrial, commercial, mining, and domestic sewage."

   (A) Solid waste (B) Hazardous waste
   (C) Non-hazardous waste (D) Both (A) and (C)

2. Which among the following is an examples of solid waste ___________

   (A) Domestic Trash (B) Scrap and empty containers
   (C) Fly-ash (D) All the above

3. ________ means all decomposing and non-decomposing combustible and non-combustible solid wastes.

   (A) Municipal Solid Waste (MSW)
   (B) Hazardous waste
   (C) Refuse
   (D) Composting

4. Following is an examples of refuse_________

   (A) Cigarette and cigar butts
   (B) Household items
   (C) Only (A)
   (D) Both (A) and (B)

5. ___________ type of waste consists of everyday items such as product packaging, grass clippings, cloths, food scraps, newspapers, paint and batteries.

   (A) Solid waste
   (B) Hazardous waste
   (C) Non-hazardous waste
   (D) Municipal Solid Waste (MSW)

6. Municipal Solid Waste (MSW) is also known as ________

(A) Trash (B) Garbage
(C) Only (A) (D) Both (A) and (B)

7. What are characteristics of Municipal Solid Waste (MSW) management?

(A) Planning (B) Designing
(C) Operating and Upgrading (D) All the above

8. Disposal of solid and hazardous waste in a non-permitted area is ________.

(A) Illegal dumping (B) Recycling
(C) Composting (D) None of the above

9. Illegal dumping of solid and hazardous waste is also known as ________

(A) Midnight dumping (B) Fly dumping
(C) Only (B) (D) Both (A) and (B)

10. Municipal Solid Waste (Management & Handling) Rules enacted in the year ________

(A) 2001 (B) 2000
(C) 2003 (D) 2002

11. Plastic Waste (Management and Handling) Rules ________

(A) 2010 (B) 2012
(C) 2011 (D) 2013

12. Bio-medical Waste (Management and Handling) Rules, ________ and draft ________

(A) 2009 and 2011 (B) 2008 and 2011
(C) 2010 and 2011 (D) 2007 and 2011

13. E-Waste (Management and Handling Rules) ________

(A) 2011 (B) 2012
(C) 2013 (D) 2010

14. Battery (Management and Handling Rules) ________

(A) 1999 (B) 2000
(C) 2002 (D) 2001

15. ________ is responsible for monitoring the compliance with the MSWM plan and the MSW Rules.

(A) SPCB (B) CPCB
(C) MOUD (D) MoEFCC

16. **Illegally dumped waste primarily ________ materials that are dumped to avoid either disposal fees or the time and effort required for proper disposal.**
    (A) Solid (B) Hazardous
    (C) Non-hazardous (D) Both (A) and (C)

17. **Examples of Illegal dumped wastes ________**
    (A) Residential and Commercial waste
    (B) Bulky items and Scrap tyres
    (C) Medical waste
    (D) All the above

18. **________ is an another word for solid waste, particularly household waste.**
    (A) Residential and Commercial waste
    (B) Garbage
    (C) Both (A) and (C)
    (D) Only (A)

19. **Materials inherently dangerous in contact, handling and disposal are ________.**
    (A) Residential and Commercial wastes
    (B) Hazardous wastes
    (C) Medical wastes
    (D) All the above

20. **Residential and commercial trash and/or garbage generated by a particular municipal area is called ________**
    (A) Municipal Solid Waste
    (B) Hazardous waste
    (C) Solid waste
    (D) Non-hazardous waste

21. **________ is a process to recover a product at the of its useful life, break it down into its constituent components, and re-incorporate it into new product which has an inherent value equal to the original product.**
    (A) Recycling (B) Refuse
    (C) Upcycling (D) Resource recovery

22. **________ is a general term for solid waste materials, also called garbage or trash.**
    (A) Resource recovery (B) Refuse
    (C) Source reduction (D) Upcycling

23. __________ is the process of obtaining materials or energy, particularly from solid waste.

(A) Resource recovery (B) Rubbish
(C) Refuse (D) Upcycling

24. ________ is an useless, unwanted or discarded material with insufficient liquid content to be free flowing. It may be agricultural, commercial, industrial, institutional, municipal, or residential in nature.

(A) Residential and Commercial waste
(B) Solid waste
(C) Medical waste
(D) Hazardous waste

25. __________ are materials considered worthless, unnecessary or offensive that is usually thrown away.

(A) Residential and Commercial wastes
(B) Trashes
(C) Municipal wastes
(D) All the above

26. __________ is a process of turning waste into more valuable products.

(A) Upcycling (B) Recycling
(C) Recovery (D) Reuse

27. Red worms are used in __________ to prepare compost from organic waste.

(A) Farm yard manure (B) Bio-fertilizers
(C) Composting (D) Vermicomposting

28. __________ decomposes organic waste, such as food scraps and yard trimmings, with microorganisms (mainly bacteria and fungi), producing a humus-like substance.

(A) Farm yard manure (B) Bio-fertilizers
(C) Composting (D) Vermicomposting

29. Optimum C : N ratios for composting Municipal Solid Waste are __________ to 40:1.

(A) 25:1 (B) 26:1 (C) 26:2 (D) 25:2

30. The key parameters controlling the composting process beside C:N ratio are __________

(A) Moisture and temperature (B) Temperature alone
(C) Oxygen (D) Both (A) and (C)

31. What are the two most important elements in the composting process?

(A) Carbon and Oxygen (B) Carbon and Potassium

(C) Carbon and Phosphorus (D) Carbon and Nitrogen

32. Which among the following are composting techniques?

(A) Windrows (B) Aerated Static pile

(C) Pre and Post-Processing (D) All the above

33. Placement of wastes into the land under controlled conditions to minimize their migration or effect on the surrounding environment is ________.

(A) Combustion (B) Incineration

(C) Landfilling (D) None of the above

34. The process of burning waste under controlled conditions in the presence of oxygen is ________.

(A) Combustion (B) Incineration

(C) Pyrolysis (D) None of the above

35. The three R's of solid waste management are "________ Recycle and Recover."

(A) Replace (B) Repair

(C) Reduce (D) Reuse

36. The following are methods of Solid Waste disposal

(A) Incineration (B) Sanitary Landfilling

(C) Pyrolysis (D) All

37. Organic matter constitutes ________ municipal solid waste in India.

(A) 35-40% (B) 35-45% (C) 45-50% (D) 45-55%

38. ________ is a form of incineration that chemically decomposes organic materials by heat in the absence of oxygen.

(A) Pelletisation (B) Cofiring

(C) Pyrolysis (D) Refuse dressing

39. Pyrolysis typically occurs under pressure and at operating temperature above ______

(A) 330°C (B) 430°C (C) 230°C (D) 530°C

40. ____________ as materials that could be discarded wastes from households but are instead reused, gifted, sold, bartered (exchanged) put out for separate collection.

(A) Waste Goods (B) Waste diverted

(C) Waste management (D) Waste pickers

41. ________ include waste related materials such as recyclables, and recyclables materials, but also certain primary products which can be substituted by recycled products.

(A) Waste pickers (B) Waste diverted

(C) Waste management (D) Waste Goods

42. ________ include human tissue, organs and body part and body fluids that are removed during surgery or autopsy or other medical procedures and specimens of body fluids and their contents.

(A) Medical Waste (B) Pharmaceutical waste

(C) Mixed waste (D) Pathological waste

43. ________ include used drugs expired date drugs, partially opened or unopened drugs, remaining in the biological laboratories attached to the hospitals.

(A) Medical Waste (B) Pharmaceutical waste

(C) Mixed waste (D) Pathological waste

44. ________ that are generated in the diagnostics treatment or immunization of human beings or animals in research pertaining or in the production or testing of biological materials.

(A) Medical Wastes (B) Pharmaceutical wastes

(C) Mixed wastes (D) Pathological wastes

45. ________ have not been separated or sorted and have generally little or non-value and it consist of all kinds of recyclables and non-recyclables materials.

(A) Composted Waste (B) Separated waste

(C) Mixed waste (D) Dumped waste

46. ________ consist of mixed waste which are dumped at the roadside or other undersigned places, causing various environment and health problems.

(A) Composted Waste (B) Separated waste

(C) Mixed waste (D) Dumped waste

47. ________ include all kinds of waste which may transmit viral, bacterial, parasitic diseases.

(A) Composted Waste (B) Separated waste

(C) Incinerated waste (D) Dumped waste

48. C:N ratio of matured compost will be around ________

(A) 12 to 14 : 1 (B) 25 to 30 : 1

(C) 18 to 25 : 1 (D) 40 to 60 : 1

49. Among Union Territories ________ generate the highest (870 g/day) per capita municipal solid waste.
    (A) Delhi
    (B) Diu and Daman
    (C) Lakshadweep Islands
    (D) Andaman and Nicobar Islands

50. Waste generation rate in Indian cities ranges between ________ grams/day, depending upon the region's lifestyle and the size of the city.
    (A) 100-670 (B) 200-870
    (C) 100-770 (D) 200-970

Answers Key

Q. Ans	Q. Ans	Q. Ans	Q. Ans	Q. Ans	Q. Ans	Q. Ans
1. D	2. D	3. C	4. D	5. D	6. D	7. D
8. A	9. D	10. B	11. C	12. B	13. A	14. D
15. A	16. C	17. D	18. B	19. B	20. A	21. A
22. B	23. A	24. B	25. B	26. A	27. D	28. C
29. A	30. D	31. D	32. D	33. C	34. B	35. C
36. D	37. A	38. C	39. B	40. B	41. D	42. D
43. B	44. A	45. C	46. D	47. C	48. A	49. D
50. B						

◈

CHAPTER 13

CLIMATE CHANGE

1. Considering climate change, melting of ice sheets and glaciers causes ________
   (A) destruction of infrastructure
   (B) endangering of species
   (C) desertification
   (D) destruction of human settlements

2. Considering climate change, increased global temperatures causes ________
   (A) desertification
   (B) destruction of human settlements
   (C) destruction of infrastructure
   (D) endangering of species

3. Considering climate change, rise in water level in oceans and seas due to melted ice sheets and glaciers causes ________
   (A) desertification
   (B) destruction of human settlements
   (C) destruction of infrastructure
   (D) endangering of species

4. Long period of time without water in an area leads to ________
   (A) flood (B) drought
   (C) desertification (D) endangering

5. Human activities that causes climate change on Earth includes ________
   (A) burning of forests (B) agricultural activities
   (C) use of aerosol cans (D) all of above

6. Humans perspire when body temperature is warm and ________
   (A) precipitation is low (B) precipitation is high
   (C) humidity is high (D) humidity is low

7. Temperature can be measured as ________
   (A) degree Celsius (B) degree Fahrenheit
   (C) percentage (D) both A and B

8. Element which is used to measure how cold or hot weather is ________
   (A) wind (B) precipitation
   (C) humidity (D) temperature

9. Considering relief rain, side of mountain which stays dry is ________
   (A) hail ward side (B) leeward side
   (C) windward side (D) none of the above

10. Amount of water vapor in air is classified as ________
    (A) wind (B) precipitation
    (C) humidity (D) temperature

11. Location of polar climate region on North Pole is ________
    (A) 50.5°N (B) 40.5°N
    (C) 25.5°N (D) 66.5°N

12. Polar climate region lies between Arctic Circle and ________
    (A) Arctic Circle (B) Antarctic Circle
    (C) Caribbean Circle (D) Pacific Circle

13. Mean monthly temperature of Polar Climate region is below ________
    (A) 7°C (B) 8°C
    (C) 10°C (D) 3°C

14. Polar climate region starts from ________
    (A) Caribbean Circle (B) Pacific Circle
    (C) Arctic circle (D) Antarctic circle

15. Fifth layer of Earth's atmosphere is ________
    (A) exosphere (B) mesosphere
    (C) thermosphere (D) troposphere

16. Layer of air which surrounds Earth is called ________
    (A) season (B) atmosphere
    (C) weather (D) climate

17. Third layer of Earth's atmosphere is ________
    (A) troposphere (B) stratosphere
    (C) mesosphere (D) thermosphere

18. First layer of Earth's atmosphere is ________
    (A) troposphere (B) stratosphere
    (C) mesosphere (D) thermosphere
19. Fourth layer of Earth's atmosphere is ________
    (A) mesosphere (B) thermosphere
    (C) troposphere (D) stratosphere
20. Number of mountains in Tibetan Plateau whose height are more than 8,000 meters are ________
    (A) 20 (B) 18 (C) 14 (D) 16
21. Highest mountain of world is ________
    (A) Mount Everest (B) K2
    (C) Mount Fuji (D) Mount Vesuvius
22. Steep-walled valleys having narrow depressions and are cut by streams and rivers are called ________
    (A) C-shaped (B) O-shaped
    (C) V-shaped valleys (D) U-shaped valleys
23. Average height of Tibetan Plateau is of ________
    (A) 7200 meters (B) 5000 meters
    (C) 6500 meters (D) 7000 meters
24. Long depression which is surrounded by high grounds is classified as ________
    (A) plain (B) valley
    (C) hill (D) plateau
25. Considering layers of Earth, temperature in center of Earth can be ________
    (A) 6500°C (B) 6800°C
    (C) 5000°C (D) 7300°C
26. Submerged range of mountains which ends to southern end of Africa from Arctic is classified as ________
    (A) Mid-African Ridge (B) Mid Eurasian Ridge
    (C) Mid-Arctic Ridge (D) Mid-Atlantic Ridge
27. Volcanic activity area is classified as ________
    (A) ridge (B) drift
    (C) crust (D) mantle
28. Large masses of land on surface of Earth is classified as ________
    (A) continents (B) oceans
    (C) hemispheres (D) longitudes

29. **Great pieces on outer layer of earth are classified as ____________**

(A) oceanic plates (B) Atlantic ridges
(C) crustal plates (D) continental drifts

30. **The greenhouse effect is caused by __________**

(A) Greenhouse gases in the lower atmosphere absorbing solar radiation
(B) Greenhouse gases in the lower atmosphere absorbing radiation from the Earth's surface, and preventing much of it escaping into space
(C) Too much heat in the atmosphere
(D) Too much sunshine reaching Earth

31. **Which of these natural events affect the climate?**

(A) Thawing permafrost
(B) The path of the Earth around the sun
(C) Volcanic eruptions
(D) All of the above

32. **Which of these greenhouse gases is most abundant in the atmosphere?**

(A) Carbon dioxide (B) Methane
(C) Nitrous oxide (D) Water vapour

33. **What greenhouse gas was not present in the atmosphere in pre-industrial times (before 1750)?**

(A) Trichlorofluoromethane (B) Carbon dioxide
(C) Nitrous oxide (D) Methane

34. **Which of the following changes in climate has NOT already occurred?**

(A) 0.74°C increase in average global temperature over the last century
(B) 0.17m rise in sea level during the 20th century
(C) A decrease in the sea ice around Antarctica
(D) Fewer cold days and frosts, and more frequent heat waves over the last 50 years

35. **Which of these countries has the highest per capita carbon dioxide emissions?**

(A) United States (B) Australia
(C) Saudi Arabia (D) China

36. Which of the following activities contributes the most to carbon emissions globally?
(A) Agriculture (B) Transport
(C) Forestry (D) Energy supply

37. In research published in *Nature* scientists claim that we have a good chance of stabilizing global warming below 2°C if the world's total carbon dioxide emissions from 2000 to 2050 don't exceed 1,000 billion tonnes. How much carbon dioxide did we emit between 2000 and 2006?
(A) 1,000 billion tonnes
(B) 234 billion tonnes
(C) 146 billion tonnes
(D) 94 billion tonnes

38. Australia emitted 597 million tonnes of greenhouse gas in 2007. Through the following one could cut up to 110 million tonnes of their annual emissions
(A) Installing energy efficient lightbulbs in every home
(B) Greater energy efficiency in industry and more energy efficient buildings
(C) Recycling all recyclable waste at home and in industry
(D) Planting one million trees

39. Which of the following is NOT expected to happen in Australia as a result of global warming?
(A) Central Australia will become drier and have more frequent and severe droughts
(B) Perth will have between 29 to 38 days above 35 °C a year by 2030
(C) Canberra will have between 26 and 29 days of extreme fire danger by 2020
(D) Tasmania will become a tropical island

40. The UNFCCC Climate Change Conference (COP23)'s roadmap for 'Talanoa Dialogue', was established at which city?
(A) Bonn, Germany (B) Berlin, Germany
(C) Paris, France (D) Moscow, Russian

41. Name the MoEFCC campaign for land use science or Forest Plus?
(A) Good Wood (B) Wood Good
(C) Wood is Good (D) Good Earth

42. Union Environment ministry approved projects for climate change in which states?

(A) Rajasthan (B) Gujarat
(C) Sikkim (D) All

43. According to a study by UK scientists, plant life in which region is growing rapidly due to climate change?

(A) Arctic (B) Antarctic
(C) Greenland (D) Australia

44. A programme to assess the status of Himalayan pheasants and finches is being carried out by ________

(A) Bombay Natural History Society
(B) World Wildlife Fund
(C) PETA
(D) None of the above

45. Climate change will cause escalation in migration according to which report?

(A) Climate Change Knows No Borders
(B) Climate Change Has No Borders
(C) Climate Change With No Borders
(D) None of the above

46. Paris Agreement on Climate Change moved closer with how many countries joining the special event hosted by UN Secretary General Ban Ki-moon?

(A) 31 (B) 32
(C) 33 (D) 34

47. Which of the following causes global warming?

(A) Carbon dioxide (B) Oxygen
(C) Nitrogen (D) Hydrogen

48. How many percent of carbon dioxide increased in the atmosphere since pre-industrial times?

(A) About 10% (B) About 20%
(C) About 30% (D) About 40%

49. What is the full form of UNFCC with respect to global warming convention?

(A) United Nations Framework Convention on Climate Change
(B) United Nations Federation Convention on Climate Change
(C) United Nations Framework Center on Climate Change
(D) United Nations Federation Center on Climate Change

50. **Who measures the global warming rate?**
(A) Astrologers (B) Physicist
(C) Philosopher (D) Climatologist

51. **Which one of the following result takes place due to global warming?**
(A) Maintaining steady temperature
(B) Changes in the rainfall
(C) Pleasant environment
(D) Causing less pollution

52. **Which one of the following cause global warming?**
(A) Radiative forcing (B) Earth gravitation force
(C) Oxygen (D) Centripetal force

53. **How human activity has influenced global warming?**
(A) By planting more trees
(B) By causing changing in gravitational force
(C) By changing the radiative balance governing the Earth
(D) By protecting environment

54. **Which one of the following is the anthropogenic radiative forcing of climate?**
(A) Aerosols (B) Cement
(C) Paper (D) Glass

55. **Which one of the following land use causes global warming?**
(A) Increase in the fertility of soil
(B) Surface reflectance
(C) Forestation
(D) Adopting organic farming

56. **Changes in the composition of the atmosphere is an example of ____________**
(A) Internal forcing (B) External forcing
(C) Mid forcing (D) Gravitational forcing

57. **What is the main reason for melting of ice sheets?**
(A) Increase in the oxygen content
(B) Global warming
(C) Decrease in carbon dioxide content
(D) Noise pollution

58. **Global warming also refers to as __________**
(A) Climate change (B) Ecological change
(C) Atmosphere change (D) None of the above

59. Global warming is observed century-scale rise in the ______ of the Earth's climate system.
(A) Maximum temperature (B) Minimum temperature
(C) Average temperature (D) Temperature

60. Which of the following is not a greenhouse gas?
(A) Carbon di-oxide (B) Methane
(C) Nitrous oxide (D) Carbon monoxide

61. Which of the following is (are) the anticipated effect(s) of Global warming?
(A) Rising sea levels (B) Changing precipitation
(C) Expansion of deserts (D) All of the above

62. UNFCCC stands for ______
(A) United Nations Framework Convention on Climate Change
(B) United Nations Federation Convention on Climate Change
(C) United Nations Framework Council on Climate Change
(D) United Nations Federation Council on Climate Change

63. Economies of the following country(ies) is (are) responsible for greatest annual CO_2 emissions.
(A) USA (B) China
(C) Both (A) and (B) (D) Brazil

64. Ocean temperatures increase more slowly than land temperatures because of the larger effective ______ of the oceans.
(A) Heat capacity (B) Heat balance
(C) Heat transfer (D) Heat exchanger

65. Which of the following types of sources of energy do not produce carbon di-oxide?
(A) Wind energy (B) Geothermal energy
(C) Hydroelectric energy (D) All of the above

66. The El-Nino Phenomenon occurs in ______
(A) Atlantic Ocean (B) Pacific Ocean
(C) Indian Ocean (D) Arctic Ocean

67. Without the Earth's atmosphere, the Earth's average temperature would be well below the ______ temperature of water
(A) Boiling (B) Freezing
(C) Vaporization (D) None of the above

68. The major gas responsible for greenhouse effect is ________

(A) Water vapor (B) Carbon dioxide
(C) Methane (D) Ozone

69. Following is (are) the cause(s) for CO_2 emissions.

(A) Fossil fuel burning
(B) Changes in land use
(C) Calcination of limestone for clinker production
(D) All of the above

70. The Ozone layer lies in ________

(A) Troposphere (B) Stratosphere
(C) Ionosphere (D) Mesosphere

71. ________ is a gradual reduction in the amount of global direct irradiance at the Earth's surface.

(A) Global warming (B) Global positioning
(C) Global dimming (D) None of the above

72. ________ is the main cause of Global dimming.

(A) Aerosols (B) Biomes
(C) Acid rain (D) All of the above

73. Carbon dioxide contributes ________ percent to global warming.

(A) 10 to 15% (B) 20 to 25%
(C) 30 to 35% (D) 40 to 45%

74. A climate model is a representation of the physical, chemical and ________ processes that affect the climate system.

(A) Biochemical (B) Biological
(C) Biotic (D) None of the above

75. Climate model may be used to predict

(A) Local air movement (B) Ocean temperature
(C) Ice cover on land (D) All of the above

76. The following is the most endangered nation due to flooding from climate change.

(A) Tuvalu (B) Bangladesh
(C) Maldives (D) Antigua

77. Following is (are) the important Kyoto mechanism(s)

(A) Emission trading
(B) Clean Development Mechanism
(C) Both (A) and (B)
(D) None of the above

78. Head quarter of IPCC is located in ________

(A) Brazil (B) USA
(C) Geneva (D) France

79. IPCC means ________

(A) Indian panel on climate change
(B) International panel on climate change
(C) Inter-governmental panel on climate change
(D) Integrated panel on climate change

80. The IPCC has published ________ assessment reports reviewing the latest climate science

(A) Five (B) Eight
(C) Three (D) Four

81. IPCC is reported that main reason for climate change is ________

(A) Human activities (B) GHG
(C) Industries (D) Environmental factors

82. The climate change conference (COP20) was held in ________

(A) Brazil (B) Peru
(C) USA (D) Canada

83. Global issues due to climate change is ________

(A) Increase in mortality
(B) Increase in agricultural productivity
(C) Decrease in temperature
(D) Decrease in water temperature

84. Which year the IPCC was created________

(A) 1980 (B) 1988
(C) 1898 (D) 1890

85. In 2050, India's average annual temperature are expected to rise the range between________

(A) 0.1 °C-0.2 °C (B) 4 °C - 5 °C
(C) 1 °C -2 °C (D) 0.5 °C -2.5 °C

86. Accumulation of ________ radiation increases the global temperature.

(A) Short wave length (B) Long wave length
(C) Solar radiation (D) Medium wave length

87. The size of the Aerosol black carbon ________

(A) 2 - 5 micrometre (B) 2.5 - 10 micrometre
(C) 3 - 8 micrometre (D) 4 - 6 micrometre

88. The latest and 24th meeting of conference of parties(COP 24) was began at ______
(A) Poland (B) India
(C) USA (D) Australia

89. Which is the impact on climate change in India ______
(A) Sea level rise
(B) Changes in agricultural yield
(C) Changes in water patterns
(D) all the above

90. UNEP stands for ______
(A) United National Environment Programme
(B) United Nations Ecosystem Project
(C) United Nationalized Environment project
(D) United Nations Environment Programme

91. Paleoclimatology means ______
(A) study of past climate changes throughout earth history
(B) study of past atmospheric changes throughout earth history
(C) study of both climate and atmospheric changes
(D) none of the above

92. Who discovered paleoclimatology ______
(A) John Hardcastle
(B) John William
(C) John Johannsen
(D) Johnson

93. Dendroclimatology is the study of ______
(A) Only climate change condition during different periods
(B) Only atmospheric condition during different periods
(C) Both A and B
(D) None of the above

94. World Meteorological Organization ______
(A) Switzerland (B) Brazil
(C) America (D) Europe

95. Percentage of oxygen in atmosphere ______
(A) 21 % (B) 38% (C) 48% (D) 18 %

96. Ability of a substance to cause global warming is known as
(A) Ground water potential (B) Global warming potential
(C) Global warming process (D) Ground water process

97. Conversion of nitrate to nitrogen is known as ________
(A) Nitrification (B) Denitrification
(C) Ammonification (D) Deammonification

98. ________ atom is the principle building block of many kinds of molecules
(A) Oxygen (B) Nitrogen
(C) Carbon (D) Sulphur

99. The bond which combines C with H,O,N is ________
(A) Wanderwalls force (B) Molecular bond
(C) Covalent bond (D) All the above

100. Guana the bat excreta contains ________
(A) Potassium (B) Phosphorus
(C) Nitrogen (D) Nitrate

101. A report on Global Environment Monitoring System in 1975 was published by ________
(A) EPA (B) WHO
(C) FAO (D) UNEP

102. Environmental Clearance for developmental project is provided by ________
(A) State Pollution Control Boards
(B) Central Pollution Control Boards
(C) Union Ministry of Environment and Forests
(D) The Impact Assessment Agency of concerned state

103. Notification on Environmental Impact Assessment of Development Projects is covered under ________
(A) Environmental Impact Assessment Act
(B) Environment Protection Act
(C) The Factories Act
(D) The Environment Conservation Act

104. Mauna Loa, in Hawaii is famous for ________
(A) Botanical Garden
(B) Monitoring Sea Level Rise since 1950
(C) Biggest Collection of Mammal Fossils
(D) Continuous monitoring of atmospheric Carbon-di-oxide since 1957

105. Green Peace is an ______
   (A) International NGO
   (B) International Peace Accord
   (C) International treaty on Ocean Dumping
   (D) UN's Environment Initiative

106. The most visually striking evidence of global warming is ______
   (A) Increased precipitation along the Gulf Coast States
   (B) Highly varying temperature fluctuations felt during winter months
   (C) Rapid melting of glacial ice on nearly every continent
   (D) All of the above

107. Atmospheric nitrogen fixation example ______
   (A) Thunder storm and lightening
   (B) Tornadoes and lightening
   (C) Thunderstorm and tornadoes
   (D) All the above

108. A leading role in Earth's climate is played by ______
   (A) Soil (B) Ocean
   (C) Mountains (D) Volcanoes

109. Most of the Earth's water is found in ______
   (A) Lakes (B) Polar Ice caps
   (C) The Oceans (D) Rivers

110. Which of these has a greater effect on global sea level ______
   (A) Melting sea ice (B) Melting land ice
   (C) Both have same effect (D) None of them affects

111. What is sea surface salinity?
   (A) The measure of Tsunami strength
   (B) The light reflected off the ocean's surface
   (C) Salt concentration on the ocean's surface
   (D) All the above

112. Tides are caused by ______
   (A) Wind
   (B) Moon's gravitational pull
   (C) Gravitational pull of both moon & sun
   (D) Earthquakes

113. The highs and lows of ocean surface are known as ________

(A) Ocean surface geography
(B) Ocean surface topography
(C) Ocean surface pressure gradient
(D) Ocean surface altitude

114. The only current that makes an uninterrupted circle around the entire earth without hitting land is the ______

(A) Gulf stream
(B) Atlantic circumpolar current
(C) Japan current
(D) California current

115. First recorded use of the term 'El Nino' was in ________

(A) Switzerland (B) German
(C) Peru (D) Sweden

116. United Nations Interagency Task Force on El-Nino was established in ________

(A) 1970 (B) 1997 (C) 1985 (D) 1967

117. Which of the following is not affected by the phenomena of El-Nino?

(A) Paddy (B) Maize
(C) Tea Plantation (D) Castor

118. What causes warm than normal sea surface temperature?

(A) La Nina (B) El Nino
(C) Coriolis force (D) ITCZ

119. El Nino is a ________ word

(A) Spanish (B) German
(C) French (D) English

120. The term El Nino translates from Spanish word as ________

(A) The girl child (B) The boy child
(C) The woman (D) The man

121. The wind seed of the cyclones during El Nino Episode is ________

(A) Moderate (B) Fast
(C) Slow (D) Very fast

122. Which of the following is a cold event?

(A) El Nino (B) Southern oscillation
(C) Northern oscillation (D) La Nina

123. The word desert is derived from the latin word Desertus meaning ________
(A) To abandon (B) To clear
(C) To leave (D) To inhabitate

124. ________ refers to degradation of land in arid, semi-arid and dry sub humid areas
(A) Desertification (B) Aridification
(C) Humidfication (D) None of these

125. What are the causes of desertification?
(A) Deforestation (B) Climate change
(C) Overpopulation (D) All the above

126. Which continent is most affected by desertification?
(A) Asia (B) Africa
(C) North America (D) Europe

127. The continent where no desertification is occurred ________
(A) Asia (B) Antarctica
(C) Africa (D) South America

128. World day to combat desertification and drought ________
(A) May 17 (B) June 17
(C) May 27 (D) June 27

129. World day to combat desertification and drought was designated by ________
(A) UNGA (B) UNSC
(C) UNTC (D) UNDP

130. 2018 theme of WDCDD ________
(A) Land has true value – invest in it
(B) Protect Earth. Restore Land. Engage People
(C) Combating Land Degradation for Sustainable Agriculture
(D) Women and Desertification

131. WDCCD was designated in ______ and was observed for the first time in ______
(A) 1993, 1994 (B) 1994, 1995
(C) 1995, 1996 (D) 1996, 1997

132. The global observance of WDCD was hosted by ______
(A) Government of Colombia (B) Government of Ecuador
(C) Government of Paraguay (D) Government of Peru

133. **Which group is responsible for large amount of desertification awareness?**

(A) DCC (B) CCD
(C) UND (D) MCD

134. **The "Green Wall" is one of the major successful methods in the battle against desertification. Where is the "Green Wall" come from?**

(A) USA (B) China
(C) Russia (D) India

135. **International Year of Deserts and Desertification?**

(A) 2004 (B) 2005
(C) 2006 (D) 2007

136. **According to UNCOD estimation, a moderate desertification leads to a ______ drop in agricultural productivity**

(A) 5 – 15% (B) 10 – 25%
(C) 20 – 35% (D) 30 – 45%

137. **According to UNEP, all regions in which ratio of total annual precipitation to potential evapotranspiration (P/ET) ranging from ______ are considered vulnerable to desertification**

(A) 0.4 to 0.55 (B) 0.5 to 0.65
(C) 0.6 to 0.75 (D) 0.7 to 0.85

138. **What causes flood?**

(A) Littering (B) Earth quake
(C) Raise of sea level (D) None of the above

139. **What are the effects of flood?**

(A) Injury and loss of life
(B) Damage of buildings
(C) Both A and B
(D) Clean drinking water

140. **Which is not a characteristic of a regional flood?**

(A) an extended rise before river elevation for weeks to month
(B) Large storm systems or an extended period of rainfall
(C) A small drainage basin
(D) Floodplains covered with floodwaters

141. **Which is the following of the key factor causation of drought?**

(A) Food supply (B) Precipitation
(C) Temperature (D) Light

142. Which research centre lists three factors that cause drought?
   (A) NASA Earth Observatory
   (B) ISRO Earth Observatory
   (C) ENVIS Earth Observatory
   (D) NOAA Earth Observatory

143. Cyclones is form where ______
   (A) All the answers are correct
   (B) High pressure air pushes wind away causing a drain effect
   (C) Two air masses meet head on and roll backwards, causing a typhoon and a hurricane
   (D) Wind is drawn a low pressure area at earth 's surface, forming a spiral pattern

144. A typhoon is an intense cyclone that occurs over the ______
   (A) Pacific Ocean (B) Bahamas islands
   (C) Indian continents (D) Atlantic ocean

145. A system for measuring drought was developed by ____ in 1965
   (A) Leimer (B) Stratousberg
   (C) Carl Kepler (D) Wayne Palmer

146. PDSI means ______
   (A) Palmer Drought Severity Index
   (B) Palmer Drought Stress Index
   (C) Precipitation Drought Severity Index
   (D) Palmer Drought Severity Institute

147. Types of Drought ______
   (A) Meteorological drought (B) Agricultural drought
   (C) Hydrological drought (D) All the above

148. An Index which is related to some of the cumulative effects of a prolonged and abnormal moisture deficiency ______
   (A) Drought Index (B) Wet Index
   (C) Salinity Index (D) Solubility Index

149. Drought contingency planning must allow for the implementation of 3 kinds of measures ______
   (A) Mitigation (B) Relief
   (C) Rehabilitation (D) All the above

150. What is the most intense part of the hurricane?
   (A) Eye (B) Eye wall
   (C) Eye band (D) Rain bands

151. What is the calmest part of the hurricane?

(A) Storm surge (B) Rain bands

(C) Eye (D) Eye wall

152. What is the name of the scale used to rank hurricanes?

(A) Tropical scale (B) Richter scale

(C) Saffir Simpson scale (D) Hurricane scale

153. What is the correct order of development of hurricane?

(A) Tropical wave, Tropical depression, tropical storm, Hurricane

(B) Tropical depression, Tropical wave, Tropical storm, Hurricane

(C) Tropical wave, Tropical storm, Tropical depression, Hurricane

(D) Tropical wave, Tropical storm, Hurricane, Tropical depression

154. Hurricanes are classified into how many categories?

(A) 2 (B) 5

(C) 8 (D) 6

155. What are the most important ingredients necessary for hurricanes to form?

(A) 80° and warmer water (B) 80° and cold water

(C) 100° and warm water (D) 100° and cold water

156. A tropical cyclone in the Atlantic is referred to as a _________

(A) Flood (B) Drought

(C) Hurricane (D) Cyclone

157. What are all the Research areas for climate change?

(A) Air quality and climate change

(B) Ecosystem and climate change

(C) Both A and B

(D) None of the above

158. Expansion of CCRC __________

(A) Climate Change Research Centre

(B) Continuing Care Retirement Communities

(C) Community Care Research Center

(D) Climate Change Reconstruction Council

159. What % of climate scientists agree that human activities are causing climate change?

(A) 97 (B) 50 (C) 47 (D) 81

160. Which is the following comes under climate change organization?

(A) ADB (B) NOAA

(C) IBRD (D) NASA

161. Sea level rose by 19 cm between 1901 and ____

(A) 1950 (B) 2000

(C) 2005 (D) 2010

162. Fourth National Climate Assessment developed by

(A) U.S (B) U.K

(C) Delhi (D) Australia

163. Current news and data streams about global warming and climate change obtained from ______

(A) NASA Earth Observatory

(B) ISRO Earth Observatory

(C) ENVIS Earth Observatory

(D) NOAA Earth Observatory

164. Ultimate goal of climate change mitigation and sustainable development is to reduce _______

(A) GHG (B) Oxygen

(C) Argon (D) Nitrogen

165. CCRI has been founded with a vision to promote _______

(A) Environment (B) Eduvation

(C) Innovation (D) All the above

166. Latest research and reviews done on climate change velocity analysis by _______

(A) Germany (B) Delhi

(C) Europe (D) Indonesia

167. Methods involve in GHG monitoring of CO_2in Atmosphere involve

(A) Infrared Analyzing (B) Monometry

(C) Titrimetry (D) All the Above.

168. Satellite Monitoring of GHG involves _________

(A) Orbiting Carbon Observatory

(B) Integrated Carbon Observation System.

(C) Earth Observation Satellite System

(D) Both A and B

169. Monometry is a tool to measure ________

(A) Temperature of Atmosphere.
(B) Pressure of Atmosphere.
(C) Moisture of Atmoshphere.
(D) None of these.

170. Mauna Loa Observatory measure CO_2 in atmosphere through ________

(A) Monometry (B) Titrimetry
(C) Infrared Gas Analyzer (D) All the above

171. The glowing filament used in IR gas Analyzer is ________

(A) Nichrome (B) Tungsten
(C) Lead (D) Copper

172. The indicator used in Titimetry method is ________

(A) Methyl orange (B) Bromothymol Blue
(C) Thymolphthalein (D) Cresolphthalein

173. The Solution used in Titrimetry through which air passed for CO_2 analysis is ________

(A) Barium hydroxide (B) Sodium hydroxide
(C) Potassium dichromate (D) None of these

174. Methane gas monitoring can be done by ________

(A) Infrared analyzer
(B) Satellite monitoring
(C) Diffrential Absorption Lidar (DIAL)
(D) ACE-FTS (Atmospheric Chemistry Experiment – Fourier Transform Spectrometer)

175. The Sources of Methane emission involves ________

(A) Active Sites (B) Closed Landfill Sites
(C) Bare lands (D) Both A and B

176. Nitrous Oxide (N_2O) monitoring can be done by ________

(A) Atmospheric Chemistry Experiment – Fourier Transform Spectrometer (ACE-FTS).
(B) Diffrential Absorption Lidar (DIAL)
(C) Satellite Monitoring System
(D) All the above

177. Satellite used to measure N_2O in Troposphere is ________

(A) SCI-SAT (B) RADAR SAT
(C) ODIN SAT (D) EDU SAT

178. Sequestration of CO_2 by living marine ecosystem is called ________

(A) Black Carbon (B) Marine Carbon
(C) Bio-Carbon (D) Blue Carbon

179. The Emission of CFCs can be monitored by ________

(A) FTIR Analyzer (Fourier Transform Infrared Spectroscopy)
(B) DIAL
(C) Infrared Analyzer
(D) None of these

180. The Programme started by WMO (World Meteorological Organization) to monitor GHG in Atmosphere is ________

(A) GHG monitoring program
(B) Global Atmosphere Watch (GAW)
(C) Integrated Atmospheric monitoring program
(D) All the above

181. The advisory group organized by GAW is ________

(A) Scientific Advisory Group for GHG (SAG – GHG)
(B) Climate Change Advisory Group
(C) Advisory group on Environment emergencies
(D) None of these

182. Last Meeting of SAG – GHG took place at ________

(A) Dallas, USA (B) La Jolla, USA
(C) Tianjin, China (D) Ningbo, China

183. The monitored data were collected and distributed by

(A) WDCGG (World Data Centre for GHG)
(B) WMO
(C) Bureau of Meteorology
(D) International Meteorological organization

184. Accumulation of ________ rays causes the greenhouse effect.

(A) Short wavelength (B) Long wavelength
(C) Both A and B (D) None of the above

185. Which among the following gas has the High Global Warming potential?

(A) CO_2 (B) CH_4 (C) SF_6 (D) N_2O

186. What is the Global warming potential of N_2O?

(A) 26 to 29 (B) 265 to 298
(C) 320 to 340 (D) 192 to 220

187. Which plants are most responsive to atmospheric CO_2 enrichment?

(A) C_3 (B) C_4

(C) Both A and B (D) None of the above

188. Term related to Plant response with respect to CO_2 enrichment ______

(A) OTC – Open top chamber (B) Poly tunnel (Poly house)

(C) Greenhouse (D) None of the above

189. Which is the most abundant greenhouse gas?

(A) CO_2 (B) Ozone

(C) Fluorinated gases (D) Water vapour

190. Size of the black carbon ______ micrometer

(A) 1.2 to 7.1 (B) 2.5 to 10

(C) 5.2 to 12.5 (D) 0.5 to 1

191. Size of the respire able suspended particulates (micrometer) ______

(A) 1.5 (B) 5.2

(C) 2.5 (D) 7.5

192. What is the layer of atmosphere that contains green house gases?

(A) Troposphere (B) Mesosphere

(C) Thermosphere (D) None of the above

193. What are the most major green house gases produced by people?

(A) Methane (B) Nitrous oxide

(C) Carbon dioxide (D) SF_6

194. Concentration of carbon dioxide before the beginning of industrial era?

(A) 300 ppm (B) 380 ppm

(C) 280 ppm (D) 310 ppm

195. The green house effect invented by ______

(A) Joseph Fourier (B) Charles

(C) Voeux (D) Geiger

196. A report on global environmental monitoring system was published in ______

(A) 1991 (B) 1973

(C) 1983 (D) 1980

197. NAPCC
(A) National action plan on climate change
(B) National action plan on cyclone change
(C) National additional plan on climate change
(D) National action plan on climate co-operation

198. NSM
(A) New York Solar Mission
(B) National sunlight Mission
(C) National Solar Mission
(D) National Sunbird Mission

199. NMEEE
(A) National Mission for Enhanced Environmental Efficiency
(B) National Mission for Enhanced Energy Efficiency
(C) National Mission for Enhanced Energy Envelop
(D) New Delhi Mission for Enhanced Energy Efficiency

200. NMSH
(A) National Mission on sub-himalayan Habitat
(B) National Mission on Siberian Habitat
(C) National Mission on Sustained Habitat
(D) National Mission on Sustainable Habitat

201. NWM
(A) National Water Mission
(B) National Waste Management
(C) National Weather Mission
(D) National Wildlife Mission

202. NMSHE
(A) National Mission for Sustaining the Hills Ecosystem
(B) National management for Sustaining the Himalayan Ecosystem
(C) National Mission for Sustaining the Himalayan Ecosystem
(D) National Mission for Sustaining the Himalayan Ecology

203. NMGI
(A) National Mission for a Growing India
(B) National Missile for a Green India
(C) National Mission for a Green India
(D) National Management for a Green India

204. NMSA
   (A) National Mission for Sustained Agriculture
   (B) National Mission for Sustainable Agriculture
   (C) National Mission for Suitable Agriculture
   (D) National Mission for Subsistence Agriculture

205. NMSKCC
   (A) National Mission on Strategic Knowledge for Cyclone Change
   (B) National Mission on Strategic Knowledge for Climate Cause
   (C) National Mission on Strategic Knowledge for Climate Change
   (D) National Management on Strategic Knowledge for Climate Change

206. How fast is sea level rising?
   (A) 0.19 inches per year (B) 0.12 inches per year
   (C) 0.11 inches per year (D) 0.15 inches per year

207. Types of adaptation
   1) Incremental, Transformational
   2) Anticipatory, Proactive
   3) Private adaptation
   4) Public adaptation
   (A) 1,2,3 (B) 1,2,4
   (C) 1,2,3,4 (D) 3,4

208. The environmental act enacted in ____
   (A) 1998 (B) 1991 (C) 1987 (D) 1986

209. Which rule support the statement –"Mammals from colder climate tend to have shorter ears and limps" ?
   (A) Bergmann's rule (B) Golger's rule
   (C) Allen's rule (D) Rench's rule

210. Central Board for Prevention and Control of Water Pollution (CBPCWP)enacted in ____
   (A) 1986 (B) 1998
   (C) 1974 (D) 1988

211. What is the amount of solid waste generated per capita in India?
   (A) 1.33 Kg per d (B) 0.4 Kg per d
   (C) 0.6 Kg per d (D) 1.98 Kg per d

212. Who coined the word smog?

(A) Haekal (B) Odem
(C) Tansley (D) Henry Des Voesx

213. How long can the members of state pollution control Board except the member secretary hold the office ?

(A) Two years (B) Three years
(C) Five years (D) Four years

214. According to WHO out of top 10 most polluting world cities nine are in ______

(A) India (B) Brazil (C) Russia (D) Iran

215. Abbreviated form of NCAP is ______

(A) National Clean Air Program
(B) National City Air Program
(C) National Carbon Air Program
(D) National Clean Airway Program

216. Which country has proposed the Green New Deal (GND) to address both climate change and economic inequality?

(A) China (B) US (C) Russia (D) Brazil

217. According to IPCC, to keep global warming below 2°C, emission of CO_2 and other Green House Gases must be halved by ______

(A) 2050 (B) 2049 (C) 2048 (D) 2047

218. ______ is the first step towards achieving more substantial global emission reductions.

(A) Montreal Protocol (B) Kyoto Protocol
(C) Earth Summit (D) Paris Treaty

219. The Agreement "substantial global emission reductions" was adopted at:

(A) Earth Summit (B) UN Climate Conference
(C) Both A and B (D) None of the above

220. The Durben Platform for Enhanced action, adopted at the ______

(A) Earth summit (B) UN conference
(C) Kyoto protocol (D) None of these

221. UN conference in 2011 was held at ______

(A) Canada (B) England
(C) India (D) South Africa

222. The EU Climate and energy package was adopted in ____
(A) 2010 (B) 2011 (C) 2009 (D) 2008

223. National Climate Change Adaptation Strategies was prepared by ________ EEA member countries
(A) 14 (B) 15 (C) 16 (D) 17

224. The European Climate Adaptation Platform (Climate - ADAPT), was launched by ____
(A) European Commission (B) EEA
(C) Both A & B (D) UN

225. The European Climate Adaptation Platform was launched in the year ____
(A) 2000 (B) 2010 (C) 2020 (D) 2012

226. Increase in global temperatures causes ________
(A) Desertification (B) Melting of ice
(C) Sea level increase (D) All

227. Human activities that causes climate change on Earth includes ________
(A) burning of forests (B) agricultural activities
(C) burning of plastics (D) all of above

228. Humans perspire when body temperature is ________
(A) warm and precipitation is low
(B) warm and precipitation is high
(C) warm and humidity is high
(D) warm and humidity is low

229. Relief rain is also known as ________
(A) Acid rain (B) Summer shower
(C) Orographic rainfall (D) none of the above

230. Which among the following is not considered as type of rain fall?
(A) Relief rainfall (B) Frontal rainfall
(C) Convectional rainfall (D) Acid rain

231. The earthy scent produced when rain falls on dry soil is ____
(A) Petriplates (B) Petridish
(C) Petrichor (D) None

232. Layer separates Troposphere and Stratosphere is ____
(A) exopause (B) mesopause
(C) thermosphere (D) tropopause

233. An area of fairly level high ground is ________

(A) plain (B) valley
(C) hill (D) plateau

234. Temperature of the surface of sun is about ________

(A) 6500 °C (B) 6800 °C
(C) 5600 °C (D) 7300 °C

235. The Moon orbits the Earth once every ________

(A) 30 days (B) 27 days
(C) 25 days (D) 31 days

236. Large masses of land on surface of Earth is ________

(A) continents (B) oceans
(C) hemispheres (D) longitudes

237. Greenhouse gases in the lower atmosphere absorbing radiation from the Earth's surface, and preventing much of it escaping into space is ________

(A) Greenhouse gases (B) Greenhouse effect
(C) Hot atmosphere (D) Heating Islands

238. Which of the following events affect the climate?

(A) Thawing permafrost
(B) The path of the Earth around the sun
(C) Volcanic eruptions
(D) All of the above

239. Most abundant GHG in the atmosphere ________

(A) Carbon dioxide (B) Methane
(C) Nitrous oxide (D) Water vapour

240. Pre-industrial times ________

(A) before 1750 (B) after 1750
(C) before 750 (D) after 750

241. Which of the following changes in climate has NOT already occurred?

(A) 0.74°C increase in average global temperature over the last century
(B) 0.17m rise in sea level during the 20th century
(C) Decrease in the sea ice around Antarctica
(D) Fewer cold days and frosts, and more frequent heat waves over the last 50 years

242. **Which of these countries has the highest per capita carbon dioxide emissions?**
(A) United States (B) Australia
(C) Saudi Arabia (D) China

243. **Which of the following activities contributes the most to carbon emissions globally?**
(A) Agriculture (B) Transport
(C) Forestry (D) Energy supply

244. **COP 23's roadmap for 'Talanoa Dialogue', was established at**
(A) Berlin, Germany (B) Paris, France
(C) Moscow, Russian (D) Bonn, Germany

245. **People for the Ethical treatment of animals is denoted as ____________**
(A) WWF (B) PETA
(C) PETTA (D) None

246. **Find the correct statement**
(A) Climate Change Knows No Borders
(B) Climate Change Has No Borders
(C) Climate Change With No Borders
(D) None of the above

247. **Paris Agreement on Climate Change moved closer with how many countries joining the special event hosted by UN Secretary __________?**
(A) General Ban Ki-moon (B) General Ban Ke-moon
(C) General Ban Ki-roon (D) General Ban Ki-toon

248. **Which one of the following NOT causes global warming?**
(A) Carbon dioxide (B) Oxygen
(C) Methane (D) Water vapour

249. **Percentage of carbon dioxide increase in the atmosphere since pre-industrial times is _____**
(A) About 20% (B) About 30%
(C) About 40% (D) About 50 %

250. **Full form of UNFCC**
(A) United Nations Framework Convention on Climate Change
(B) United Nations Federation Convention on Climate Change
(C) United Nations Framework Center on Climate Change
(D) United Nations Federation Center on Climate Change

251. Global warming rate is measured by ________

(A) Astrologers (B) Physicist
(C) Philosopher (D) Climatologist

252. Which one of the following cause global dimming?

(A) Carbon di oxide (B) Earth gravitation force
(C) Black carbon (D) Moon

253. Following is the anthropogenic radiative forcing of climate

(A) Aerosols (B) Cement
(C) Paper (D) Glass

254. Which one of the following land use causes global warming?

(A) Increase in the fertility of soil
(B) Surface reflectance
(C) Forestation
(D) Adopting organic farming

255. Changes in the composition of the atmosphere is an example of________

(A) Internal forcing (B) External forcing
(C) Mid forcing (D) Gravitational forcing

256. Global warming is observed century-scale rise in the ______ of the Earth's climate system.

(A) Maximum temperature (B) Minimum temperature
(C) Average temperature (D) Temperature

257. Which of the following is not a greenhouse gas?

(A) Carbon di-oxide (B) Methane
(C) Water vapour (D) Carbon monoxide

258. The anticipated effect of Global warming

(A) Rising sea levels (B) Changing precipitation
(C) Expansion of deserts (D) All of the above

259. EPA stands for ________

(A) Environmental Preservation Agency
(B) Environmental Protection Agent
(C) Environmental Protection Agency
(D) Environmental Protein Agent

260. Which of the following types of sources of energy do not produce carbon di-oxide?

(A) Wind energy (B) Geothermal energy
(C) Hydroelectric energy (D) All of the above

261. The El-Nino Phenomenon connected with ______
(A) Atlantic Ocean (B) Pacific Ocean
(C) Indian Ocean (D) Arctic Ocean

262. Without the Earth's atmosphere, the Earth's average temperature would be well below the freezing temperature of ______
(A) Oil (B) Water
(C) Both A and B (D) None of the above

263. The major gas responsible for greenhouse effect is ______
(A) Water vapor (B) Carbon dioxide
(C) Methane (D) Ozone

264. The Ozone layer lies in ______
(A) Troposphere (B) Stratosphere
(C) Ionosphere (D) Mesosphere

265. ______ is a gradual reduction in the amount of global direct irradiance at the Earth's surface.
(A) Global warming (B) Global positioning
(C) Global dimming (D) None of the above

266. The following is the most endangered nation due to flooding from climate change.
(A) Tuvalu (B) Bangladesh
(C) Maldives (D) Antigua

267. Following is (are) the important Kyoto mechanism(s)
(A) Emission trading
(B) Clean Development Mechanism
(C) Both (A) and (B)
(D) None of the above

268. Head quarters of IPCC is located in ______
(A) Brazil (B) USA
(C) Geneva (D) France

269. The climate change conference (COP 20) was held in ______
(A) Brazil (B) Peru
(C) USA (D) Canada

270. UNEP stands for ______
(A) United National Environment Programme
(B) United Nations Ecosystem Project
(C) United Nationalized Environment project
(D) United Nations Environment Programme

271. Paleoclimatology means ________
(A) study of past climate changes throughout earth history
(B) study of past atmospheric changes throughout earth history
(C) study of both climate and atmospheric changes
(D) none of the above

272. World Meteorological Organization is in ________
(A) Switzerland (B) Brazil
(C) America (D) Europe

273. Mauna Loa, in Hawaii is famous for ________
(A) Botanical Garden
(B) Monitoring Sea Level Rise since 1950
(C) Biggest Collection of Mammal Fossils
(D) Continuous monitoring of atmospheric Carbon-di-oxide since 1957

274. Which is the first mammal in the world to be extinct due to climate change?
(A) Australian rat (B) Desert rat kangaroo
(C) African bulldog (D) None of these

275. Which one of these is not GHG?
(A) CO_2 (B) CH_3 (C) N_2O (D) Water vapor

276. Global warming potential of methane?
(A) 28–30% (B) 35-46% (C) 44% (D) 51%

277. Which one of the following has highest GWP?
(A) CH_4 (B) N_2O (C) SF_6 (D) Water vapour

278. Ultimate goal of climate change mitigation and sustainable development is to reduce
(A) GHG (B) Oxygen (C) Argon (D) Nitrogen

279. Which of the following statements is WRONG regarding the global carbon cycle?
(A) The carbon dioxide in the atmosphere is the largest storage pool of carbon.
(B) Carbonate rocks deep in the earth's crust represent the largest storage pool of soil carbon.
(C) The amount of carbon in the atmosphere is approximately 1000 times more than the amount stored in carbonate rocks.
(D) There is about one-third as much carbon in the soil as there is in the atmosphere.

280. The C/N ratio of organic materials tends to _____ during the composting process.
(A) Decline (B) Increase
(C) Remain constant (D) Fluctuate in cycles

281. COP denotes
(A) Conference of Parties (B) Conference on Parties
(C) Conference of Paris (D) Conference of Pollution

282. COP 25 held at
(A) France (B) India
(C) USA (D) Spain

283. Humans perspire when body temperature is warm and
(A) precipitation is low (B) precipitation is high
(C) humidity is high (D) humidity is low

284. Fifth layer of Earth's atmosphere is
(A) exosphere (B) mesosphere
(C) thermosphere (D) troposphere

285. Considering layers of Earth, temperature in center of Earth can be
(A) 6500°C (B) 6800°C (C) 5000°C (D) 7300°C

286. Third layer of Earth's atmosphere is
(A) exosphere (B) mesosphere
(C) thermosphere (D) troposphere

287. Transfer of atmospheric CO_2 into stable long lived carbon pool is known as
(A) Carbon conservation (B) Carbon capture
(C) Carbon sequestration (D) Carbon preservation

288. Which among the following are not passive carbon pools in soil?
(A) Humic acid (B) Fulvic acid
(C) Humin (D) Carbonic acid

289. Glycoprotein substance glomalin is produced by
(A) Bacteria (B) Algae
(C) Actinomycetes (D) Mycorrhizae

290. Pyrolysis of Biomass, (450 to 750°C) in the absence of oxygen produces a pure form of carbon is known as
(A) Biomass (B) Black carbon
(C) Biochar (D) Bio gold

291. The acidification of oceans is increasing. Why is this phenomenon a cause of concern?

a) The growth and survival of calcareous phytoplankton will be adversely affected.

b) The growth and survival of coral reefs will be adversely affected.

c) The survival of some animals that have phytoplanktonic larvae will be adversely affected.

d) The cloud seeding and formation of clouds will be adversely affected.

Which of the statements given above is/are correct?

(A) a, b, and c are correct (B) b is correct

(C) a and c are correct (D) all the statements are correct

292. Ocean acidification describes the decrease in ocean pH that is caused by ________ emissions into the atmosphere.

(A) Anthropogenic carbon dioxide (CO_2)

(A) Methane

(C) Nitrous oxide

(D) Ozone

293. Changes in composition of atmosphere is an example of ________

(A) Internal forcing (B) External forcing

(C) Mid forcing (D) Gravitational forcing

294. Which one of the following is the most anthropogenic radioactive forcing of climate?

(A) Aerosols (B) Cement

(C) Paper (D) Glass

295. The term UNFCC means ________

(A) United Nations Framework Convention on Climate Change

(B) United Nations Federation Convention on Climate Change

(C) United Nations Framework Center on Climate Change

(D) United Nations Federation Center on Climate Change

296. Satellite Monitoring of GHG involves ________

(A) Orbiting Carbon Observatory

(B) Integrated Carbon Observation System.

(C) Earth Observation Satellite System

(D) Both A and B

297. Monometry is a tool to measure ________
(A) Temperature of Atmosphere
(B) Pressure of Atmosphere
(C) Moisture of Atmosphere
(D) None of these

298. Mauna Loa Observatory measure CO_2 in atmosphere through ________
(A) Monometry (B) Titrimetry
(C) Infrared Gas Analyzer (D) All the above

299. Methane gas monitoring can be done by ________
(A) Infrared analyzer
(B) Satellite monitoring
(C) Differential Absorption Lidar (DIAL)
(D) ACE-FTS (Atmospheric Chemistry Experiment – Fourier Transform Spectrometer)

300. Accumulation of ________ rays causes the greenhouse effect.
(A) Short wavelength (B) Long wavelength
(C) Both A and B (D) None of the above

301. Term related to Plant response with respect to CO_2 enrichment is
(A) OTC – Open top chamber (B) Poly tunnel (Poly house)
(C) Greenhouse (D) None of the above

302. NSM denotes
(A) New York Solar Mission (B) National sunlight Mission
(C) National Solar Mission (D) National Sunbird Mission

303. NMEEE
(A) National Mission for Enhanced Environmental Efficiency
(B) National Mission for Enhanced Energy Efficiency
(C) National Mission for Enhanced Energy Envelop
(D) New Delhi Mission for Enhanced Energy Efficiency

304. ________ is the first step towards achieving more substantial global emission reductions.
(A) Montreal Protocol (B) Kyoto Protocol
(C) Earth Summit (D) Paris Treaty

305. A leading role in Earth's climate is played by ________
(A) Soil (B) Ocean
(C) Mountains (D) Volcanoes

306. Tides are caused by ____________

(A) Wind

(B) Moon's gravitational pull

(C) Gravitational pull of both moon & sun

(D) Earthquakes

307. First recorded use of the term 'El Nino' was in ____________

(A) Switzerland (B) German

(C) Peru (D) Sweden

308. The term El Nino translates from Spanish word ______

(A) the girl child (B) the boy child

(C) the woman (D) the man

309. Which of the following is a cold event?

(A) El Nino (B) Southern oscillation

(C) Northern oscillation (D) La Nina

310. The word desert is derived from the latin word Desertus meaning ______

(A) To abandon (B) To clear

(C) To leave (D) To inhabitate

311. Which among the following is considered as an important strategy for agriculture to reduce GHG emission

(A) Conservation Emission (B) Climate Smart Agriculture

(C) Carbon Sequestration (D) All

312. Which among the following is not a nitrification inhibitors?

(A) Neem oil (B) Karanja seed extract

(C) Nitrophyris (D) Biochar

313. The technology which is environmental friendly, developed and used in such a way so that it doesn't disturb our environment and conserve natural resources are known as

(A) Green technology (B) Environmental technology

(C) Clean technology (D) All

314. Airtight bag used for collecting gas samples (CH_4 or N_2O) from static closed chamber is __________

(A) Tedlar bags (B) Polythene bags - G9

(C) Polythene bags - G8 (D) Polythene bags - G10

315. Which among the following instrument is used to measure CH_4?

(A) GC (B) HPLC (C) AAS (D) FTIR

316. Concentration of methane is estimated chromatographic fitted with
(A) Flame Ignition Detector (B) Flame Ionization Detector
(C) Electron Capture Detector (D) Flame Detector

317. The concentration of nitrous oxide is gas sample is estimated using gas chromatographic fitted with
(A) Flame Ignition Detector
(B) Flame Ionization Detector
(C) Electron Capture Detector
(D) Flame Detector

318. The following methods are used for the quantitative analysis of CO_2
1) Static closed chamber method
2) Alkali trap method
3) Soil respirator method
4) Infrared gas analyzer method
(A) 1 & 2 are correct (B) 2 & 3 are correct
(C) 3 & 4 are correct (D) All the four (1,2,3,4) are correct

319. Which one of the following cause global warming?
(A) Radiative forcing (B) Earth gravitation force
(C) Oxygen (D) Centripetal force

320. Which one of the following is the anthropogenic radiative forcing of climate?
(A) Aerosols (B) Cement
(C) Paper (D) Glass

321. Which one of the following land use causes global warming?
(A) Increase in the fertility of soil
(B) Surface reflectance
(C) Forestation
(D) Adopting organic farming

322. Changes in the composition of the atmosphere is an example of________________
(A) Internal forcing (B) External forcing
(C) Mid forcing (D) Gravitational forcing

323. Which of the following is not a greenhouse gas?
(A) Carbon di-oxide (B) Methane
(C) Nitrous oxide (D) Carbon monoxide

324. Which of the following is (are) the anticipated effect(s) of Global warming?
    (A) Rising sea levels (B) Changing precipitation
    (C) Expansion of deserts (D) All of the above

325. Which of the following types of sources of energy do not produce carbon di-oxide?
    (A) Wind energy (B) Geothermal energy
    (C) Hydroelectric energy (D) All of the above

326. The El-Nino Phenomenon occurs in ______
    (A) Atlantic Ocean (B) Pacific Ocean
    (C) Indian Ocean (D) Arctic Ocean

327. The Ozone layer lies in ______
    (A) Troposphere (B) Stratosphere
    (C) Ionosphere (D) Mesosphere

328. ______ is a gradual reduction in the amount of global direct irradiance at the Earth's surface.
    (A) Global warming (B) Global positioning
    (C) Global dimming (D) None of the above

329. The following is the most endangered nation due to flooding from climate change.
    (A) Tuvalu (B) Bangladesh
    (C) Maldives (D) Antigua

330. Following is (are) the important Kyoto mechanism(s)
    (A) Emission trading
    (B) Clean Development Mechanism
    (C) Both (A) and (B)
    (D) None of the above

331. Head quarter of IPCC is located in ______
    (A) Brazil (B) USA
    (C) Geneva (D) France

332. The climate change conference (COP 20) was held in ______
    (A) Brazil (B) Peru (C) USA (D) Canada

333. UNEP stands for ______
    (A) United National Environment Programme
    (B) United Nations Ecosystem Project
    (C) United Nationalized Environment project
    (D) United Nations Environment Programme

334. Paleoclimatology means ________
   (A) study of past climate changes throughout earth history
   (B) study of past atmospheric changes throughout earth history
   (C) study of both climate and atmospheric changes
   (D) none of the above

335. World Meteorological Organization is in ________
   (A) Switzerland (B) Brazil
   (C) America (D) Europe

336. Mauna Loa, in Hawaii is famous for
   (A) Botanical Garden
   (B) Monitoring Sea Level Rise since 1950
   (C) Biggest Collection of Mammal Fossils
   (D) Continuous monitoring of atmospheric Carbon-di-oxide since 1957

337. Which is the first mammal in the world to be extinct due to climate change?
   (A) Australian rat (B) Desert rat kangaroo
   (C) African bulldog (D) None of these

338. Which one of these is not GHG?
   (A) CO_2 (B) CH_3 (C) N_2O (D) Water vapor

339. Global warming potential of N_2O?
   (A) 28-30% (B) 35-46% (C) 265 to 298 (D) 51%

340. Which one of the following has highest GWP?
   (A) CH_4 (B) N_2O (C) SF_6 (D) Water vapour

341. The acidification of oceans is increasing. Why is this phenomenon a cause of concern?
   a. The growth and survival of calcareous phytoplankton will be adversely affected.
   b. The growth and survival of coral reefs will be adversely affected.
   c. The survival of some animals that have phytoplanktonic larvae will be adversely affected.
   d. The cloud seeding and formation of clouds will be adversely affected.

   Which of the statements given above is/are correct?
   (A) a, b, and c are correct (B) b is correct
   (C) a and c are correct (D) all the statements are correct

342. Ocean acidification describes the decrease in ocean pH that is caused by ________ emissions into the atmosphere.
(A) Anthropogenic carbon dioxide (CO_2)
(B) Methane
(C) Nitrous oxide
(D) Ozone

343. Changes in composition of atmosphere is an example of ________
(A) Internal forcing (B) External forcing
(C) Mid forcing (D) Gravitational forcing

344. Which one of the following is the most anthropogenic radioactive forcing of climate?
(A) Aerosols (B) Cement
(C) Paper (D) Glass

345. Satellite Monitoring of GHG involves ________
(A) Orbiting Carbon Observatory
(B) Integrated Carbon Observation System.
(C) Earth Observation Satellite System
(D) Both A and B

346. Monometry is a tool to measure ________
(A) Temperature of Atmosphere
(B) Pressure of Atmosphere
(C) Moisture of Atmosphere
(D) None of these

347. Mauna Loa Observatory measure CO_2 in atmosphere through ________
(A) Monometry (B) Titrimetry
(C) Infrared Gas Analyzer (D) All the above

348. Methane gas monitoring can be done by ________
(A) Infrared analyzer
(B) Satellite monitoring
(C) Differential Absorption Lidar (DIAL)
(D) ACE-FTS (Atmospheric Chemistry Experiment – Fourier Transform Spectrometer)

349. Accumulation of ________ rays causes the greenhouse effect.
(A) Short wavelength (B) Long wavelength
(C) Both A and B (D) None of the above

350. The term La Nino translates from Spanish word ____

(A) the Little girl
(B) the boy child
(C) the woman
(D) the man

351. The C/N ratio of organic materials tends to ____ during the composting process.

(A) Decline
(B) Increase
(C) Remain constant
(D) Fluctuate in cycles

352. Anaerobic condition favours high emission of ____

(A) CO_2 (B) O_3 (C) CH_4 (D) O_2

353. Which among the following is the warmest 30 year period?

(A) 1983 to 2012
(B) 1950 to 1979
(C) 1900 to 1929
(D) 2000 to 2029

354. Global warming potential of CH_4 is ____

(A) 265 (B) 34 (C) 1 (D) 10900

355. IPCC connected with

(A) Ozone depletion
(B) Acid rain
(C) Climate change
(D) Endangered species

356. UNFCCC denoted

(A) United Nations Framework Convention on Climate Change
(B) United Nations Framework Convention on carbon content
(C) United Nations Framework Convention on on carbon capture
(D) United Nations Foundation and Convention on Climate change

357. WMO stands for

(A) Word Mountain Organization
(B) World Meteorological Organization
(C) Wider Measurement of Ozone
(D) Wind Mill Orbiter

358. Which among the following is not connected with Milankovitch Theory?

(A) Eccentricity
(B) Obliquity
(C) Procession
(D) Eclipse

359. Year without summer due to volcanic emption of Indonesia's mount Tambora

(A) 1716
(B) 1816
(C) 1916
(D) 2016

360. Meaning of El Nino in Spanish

(A) Boy child (B) Girl child
(C) Little child (D) Fat boy

361. ENSO is

(A) Southern Oscillation (B) Northern Oscillation
(C) Easter Oscillation (D) Western Oscillation

362. Ocean connected with El Nino

(A) Pacific (B) Atlantic
(C) Indian (D) Arctic

363. ONI means

(A) Oceanic Nano Index (B) Oceanic Nino Index
(C) Oceanic Natural Index (D) Ozone Nino Index

364. SOI denoted

(A) Southern Oscillation Information
(B) Southern Orbital Information
(C) Southern Oscillation Index
(D) South Oscillation Information

365. Which among the following is not a climate active pollutant?

(A) CH_4 (B) CO_2
(C) N_2O (D) O_2

366. Who discovered Acid Rain?

(A) Robert Angus Smith (B) EO Wilson
(C) Joseph Fourier (D) Odum

367. SLCP denoted

(A) Single Laser Capture Potential
(B) Short Lived Climate Pollutant
(C) Short Lived Climate Problem
(D) Short and Long Climate Pollutant

368. Which among the following is an example of sedimentary cycle?

(A) Carbon (B) Nitrogen
(C) Sulphur (D) Oxygen

369. Conversion of NO_3 to nitrogen

(A) Nitrification (B) Denitrification
(C) Ammonification (D) Nitrogen fixation

370. Ramsar convention

(A) 1971 (B) 1981 (C) 1991 (D) 2001

371. **Earth summit**
(A) 1982 (B) 1992 (C) 2002 (D) 2012

372. **World Environment day**
(A) June 5 (B) July 5 (C) August 5 (D) May 5

373. **Kyoto protocol, 1997, conducted at**
(A) USA (B) Japan (C) India (D) Thailand

374. **CDM denoted**
(A) Clean Disposal Mechanisms
(B) Combined Defense Mechanisms
(C) Carbon Destruction Mechanisms
(D) Clean Development Mechanisms

375. **UNEP means**
(A) United Nations Environment protection
(B) United Nations Environment proposal
(C) United Nations Environment programme
(D) United Nations Environment protocol

376. **Which among the following animal live long period without water?**
(A) Camel (B) Rat
(C) Fox (D) Lion

377. **Which among the following is not connected to Tundra region**
(A) Polar bear (B) Penguins
(C) Owl (D) Aretie fox

378. **Which among the following is an example of physiological adaptation by animals when changes occur in the environment**
(A) Hibernation (B) Camouflage
(C) Fleeing (D) Hiding

379. **Father of simulation modeling __________**
(A) David (B) David Cwiertney
(C) De hit (D) De Wit

380. **Which among the following are used to simulate Indian climatic condition?**
(A) RCM (B) $HadCM_3$
(C) GCM (D) All

381. **World Day to Combat Desertification and Drought**
(A) May 17 (B) June 17
(C) May 27 (D) June 27

382. The "Green Wall" is one of the major successful methods in the battle against desertification. Where is the "Green Wall" come from?
(A) US (B) China (C) Russia (D) India

383. A typhoon is an intense cyclone that occurs over the
(A) Pacific Ocean (B) Bahama islands
(C) Indian continents (D) Atlantic ocean

384. PDSI means
(A) Palmer Drought Severity Index
(B) Palmer Drought Stress Index
(C) Precipitation Drought Severity Index
(D) Palmer Drought Severity Institute

385. What is the most intense part of the hurricane?
(A) Eye (B) Eye wall
(C) Eye band (D) Rain bands

386. Small PAHs are
(A) PAHs containing up to 6 fused aromatic rings
(B) PAHs containing up to 12 fused aromatic rings
(C) PAHs containing up to 10 fused aromatic rings
(D) PAHs containing up to 15 fused aromatic rings

387. PAHs containing more than 6 aromatic rings are known as
(A) Small PAHs (B) Medium PAHs
(C) Large PAHs (D) Tiny PAHs

388. One Dobson unit is equal to ______
(A) 0.01 mm thickness at STP (B) 0.02 mm thickness at STP
(C) 0.05 mm thickness at STP (D) 0.10 mm thickness at STP

389. Aromatic hydro carbons used commonly in home are
(A) Benzene (B) Phenanthrane
(C) Anthracene (D) Naphthalene

390. Stockholm convention on persistent organic pollutants effected world wide of PCB in the year
(A) 2001 (B) 2002 (C) 2003 (D) 2004

391. Which climate change meeting is also called as earth summit?
(A) United Nations Conference on Sustainable Development
(B) Kyoto summit
(C) UN Climate Change Conference Paris
(D) Kigali climate summit

392. Between 2015-2019, which month was considered as the hottest month on the earth?
(A) May, 2019
(B) June, 2019
(C) July, 2019
(D) None of the above

393. NDC denotes ____________. It is an executive body established by the Government of India in August 1952, which is neither a constitutional nor a statutory body. It is the apex body to take decisions on matters related to approval of five year plans of the country. Prime minister is the ex-officio chairman of the NDC.
(A) National Drug Code
(B) Nationally Determined Contributions
(C) Number of District Categories
(D) National Development Council

394. How much is the India's long term goal of increasing geographical area under forest cover
(A) 50%
(B) 33%
(C) 35%
(D) 25%

395. Which treaty acts as a framework for international cooperation to combat climate change by limiting average global temperature increases and the resulting climate change, and coping with impacts?
(A) ICERD
(B) ICCPR
(C) UNFCCC
(D) UNCCD

396. If the 2015 Paris Agreement commitment on global temperature rise is not followed, climate scientists suggest a scenario by 2050, a "tipping point" beyond which our planet's natural processes trigger uncontrollable warming. Name the scenario.
(A) Water sheet scenario
(B) Heat wave scenario
(C) Polluted air scenario
(D) Hothouse Earth scenario

397. A sea-ice-free Arctic summer is among the doomsday scenarios climate scientists predict will happen much sooner than estimated. For a sea-ice-free Arctic summer, what should be the volume of ice that the Arctic Ocean contains?
(A) 1 million sq. km
(B) 100 million sq. km
(C) 10 million sq. km
(D) 1 sq. km

398. **Why are forests important for mitigating climate change?**
   (A) Forests serve as a sink in the carbon cycle
   (B) Trees provide building materials
   (C) Trees are an important food source
   (D) Leaves of trees reflect all sunlight away from the Earth

399. **How are warming ocean temperatures impacting Fish species & other marine mammals?**
   (A) They are more satisfied because of so hot
   (B) Summer seems longer so fish are schooling less
   (C) Their migratory patterns are changing
   (D) It increasing habitat safety

400. **How much degree Celsius of temperature we should limit below preindustrial era as per recent IPCC special report**
   (A) 2° C (B) 1.5 ° C
   (C) 3 °C (D) 1 °C

401. **The International Day for Preservation of Ozone Layer (World Ozone Day) is observed every year on**
   (A) September 13 (B) September 01
   (C) September 28 (D) September 16

402. **Who popularized the term "Global warming" and which year?**
   (A) Wallace Smith Broecker in the year 1975
   (B) Svante Arrhenius in the year 1896
   (C) Charles Keeling in the 1958
   (D) Joseph Black in the year 1750

403. **The Biggest award for Environmental activist in India is given in the name ____________**
   (A) Lal Bahadur Shasthri (B) Morarji Desai
   (C) Indira Gandhi (D) Kamaraj

404. **Satellite launched by ISRO for Climate Change Studies**
   (A) CARTO SAT 2B (B) GSAT 31
   (C) INSAT 3D (D) All the above

405. **Globally, which of the following economic sectors emits the largest percentage of greenhouse gas emissions?**
   (A) Transportation (B) Building
   (C) Industry (D) Electricity and Heat

406. **How much have sea levels risen in the past 100 years?**
   (A) 6 inches (B) 3.5 inches
   (C) 2 inches (D) 22 inches

407. The world water day 2020 has its theme
(A) Water and climate Change (B) Water and weather
(C) Climate and weather (D) All the above

408. How much area of forest has been burnt in the 2019 forest fire in Australia?
(A) 56 Million Acres (B) 25 Million Acres
(C) 100 Million Acres (D) 10 Million Acres

409. What is the leading cause of coral bleaching?
(A) CO_2 dissolved in ocean water
(B) Sunscreen in the water
(C) Ocean pollution
(D) Rising water temperature

410. In which of the following ways climate change mitigation can be achieved? (1) Increasing the Capacity of carbon sinks, (2) Reforestation not Afforestation
(A) 1 only (B) 2 only
(C) Both1&2 (D) Neither 1 nor 2

411. Global study report for buildings and construction emphasizes on
(A) Zero emission (B) Efficient
(C) Resilient buildings (D) All of the above

412. In which Sustainable Development Goal deliberate CLIMATE ACTION"
(A) 11 (B) 16
(C) 13 (D) 6

413. As per India's Nationally Determined Contribution (NDC) reduction emissions intensity Percentage is
(A) 33 to 35 (B) 10 to 20
(C) 15 to 25 (D) 5 to 10

414. The term "Carbon Credit" originated from
(A) Montreal Protocol (B) Kyoto Protocol
(C) Earth Summit (D) G-8 Summit

415. Which are the two countries which ratified the Paris Agreement in February 2020?
(A) China and India
(B) Kyrgyzstan and Lebanon
(C) Afghanistan and Pakistan
(D) Finland and Netherlands

416. Implementing projects that utilize biodiversity and ecosystem services is called

(A) Biodiversity based adaptation
(B) Climate based adaptation
(C) Ecosystem based adaptation
(D) Eco services based adaptation

417. Which of the following are considered secondary pollutant?

(A) Ozone (B) Benzene
(C) Sulphur di oxide (D) Carbon monoxide oxide

418. Ecologist awarded Nobel peace prize for Green Belt Movement

(A) Malala Yousafzai (B) Greeta Thunberg
(C) Wangari Mathai (D) Theresa May

419. The second hottest year as per WMO report

(A) 2017 (B) 2019 (C) 2013 (D) 2015

420. The National Action Plan on Climate Change (NAPCC) taken up by Ministry of Environment, Forests and Climate Change (MoEF&CC), GoI clearly outlines its first principle as

(A) "Protecting the coastal ecosystem and conservation due to climate change"
(B) "Protecting the poor and vulnerable sections of the society through inclusive and sustainable development strategy, sensitive to climate change"
(C) "Enhancing Rainwater harvesting and tackle the effect of Climate change"
(D) "Engineering new and innovative forms of market, regulatory and voluntary mechanisms to promote sustainable development"

421. Which among the following is not a Short Lived Climate Forcer (SLCF)

(A) Black Carbon (B) Methane
(C) Ozone (D) Carbon Dioxide

422. How many tonnes of plastic are entering the ocean each year

(A) 2.0 million tonnes (B) 8.0 million tonnes
(C) 3.0 million tonnes (D) 0.5 million tonnes

423. The term refers for the averaging of atmospheric conditions over longer periods of time

(A) Weather (B) Climate
(C) Now casting (D) Extreme events

424. Name the Emission Scenario Used by IPCC for assessment report 5

(A) Special Report on Emissions Scenarios
(B) Representative Concentrative Pathways
(C) Shared Socioeconomic Pathways (SSPs)
(D) AIM pathways

425. What is Paleoclimatology?

(A) Statistical average of 100 years climate change data
(B) Study of Climate of the past ages
(C) Sea level change indicator
(D) None of the above

426. How many countries signed Paris agreement for Climate Change

(A) 180 (B) 198
(C) 100 (D) 200

427. What is the rank of India as per list of countries in Climate performance index released on 2019 ?

(A) 1 (B) 9
(C) 15 (D) 20

428. What percentage of the global greenhouse gas emissions does the transportation sector emit?

(A) 14% (B) 22%
(C) 5% (D) 0.5%

429. What are the Green House Gases ratifies under Kyoto protocol?

(A) CO_2, Ozone, Black carbon
(B) CO_2, Methane, Hydrofluorocarbons
(C) Ozone, CFCs, Black Carbon
(D) Nitrogen, Oxygen, O_3

430. Who was the president of the Landmark Paris COP?

(A) Carolina Schmidt (B) Michal Kurtyka
(C) Frank Bainimarama (D) Manuel Pulgar-Vidal

431. The Indian sub-continent experiences comparatively milder winters as compared to Central Asia due to which of the following factors?

(A) The Tropic of Cancer (B) The surrounding seas
(C) The Himalayas (D) Ocean currents

432. Which of the following atmospheric conditions govern the climate and associated weather conditions in India?

(A) Pressure and surface winds
(B) Upper air circulation
(C) Western cyclonic disturbances and tropical cyclones
(D) All of the above

433. What kind of weather will nimbostratus clouds likely bring?

(A) Sunny (B) Rainy
(C) Warm (D) Thunderstorms

434. What kind of weather would an anticyclone likely bring?

(A) Stormy (B) Cool and wet
(C) Dry and clear (D) Changeable

435. During El Nino period the temperature rises rapidly once in three to eight years along the coast of __________

(A) Peru and Ecuador (B) Spain and France
(C) India and Pakistan (D) China and Japan

436. Which of these animals were badly affected by Australian bush fire?

(A) Koala and Kangaroo
(B) Kangaroo and Giant panda
(C) Giant panda and Koala
(D) Kangaroo and elephant

437. Select correct statement with reference to global warming

(A) Is the short-term heating of Earth's climate system observed since the post-industrial period due to human activities
(B) Is the long-term heating of Earth's climate system observed since the pre-industrial period due to natural activities
(C) Is the long-term heating of Earth's climate system observed since the pre-industrial period due to human activities
(D) Is the short-term heating of Earth's climate system observed since the pre-industrial period due to natural activities

438. Which is the most hazardous health stressor in the dust category?

(A) PM 100
(B) PM 1
(C) PM 2.5
(D) Suspended Particulate matter

439. Identify the Government Scheme which promotes LED lights for Energy Conservation
(A) National Solar Mission
(B) Tamil Nadu Electric Vehicle Policy
(C) Ujala Yojana
(D) Jawaharlal Nehru National Solar Mission

440. Besides landforms and drainage, which is the third basic element of the natural environment?
(A) Climate (B) Temperature
(C) Wind (D) Rainfall

441. The Special Report on Emissions Scenarios (SRES) is a report by the Intergovernmental Panel on Climate Change (IPCC) that was published in
(A) 2000 (B) 2001
(C) 2002 (D) 2003

442. ______________ is a greenhouse gas concentration (not emissions) trajectory adopted by the IPCC. Four pathways were used for climate modeling and research for the IPCC fifth Assessment Report (AR5) in 2014.
(A) Special Report on Emissions Scenarios
(B) Representative Concentrative Pathways
(C) Shared Socioeconomic Pathways (SSPs)
(D) AIM pathways

443. Swedish scientist that was the first to claim in 1896 that fossil fuel combustion may eventually result in enhanced global warming. He proposed a relation between atmospheric carbon dioxide concentrations and temperature.
(A) Wallace Smith Broecker in the year 1975
(B) Svante Arrhenius in the year 1896
(C) Charles Keeling in the 1958
(D) Joseph Black in the year 1750

444. An American scientist whose recording of carbon dioxide at the Mauna Loa Observatory first alerted the world to the possibility of anthropogenic contribution to the "greenhouse effect" and global warming.
(A) Wallace Smith Broecker in the year 1975
(B) Svante Arrhenius in the year 1896
(C) Charles Keeling in the 1958
(D) Joseph Black in the year 1750

445. ______________ Began his work on the chemistry of "magnesia alba" (a basic magnesium carbonate), which he later submitted for his MD thesis in Edinburgh, and which includes the discovery of what we now call carbon dioxide - he called it "fixed air"

(A) Wallace Smith Broecker in the year 1975
(B) Svante Arrhenius in the year 1896
(C) Charles Keeling in the 1958
(D) Joseph Black in the year 1750

446. Most abundant element in sea water is __________

(A) Sodium (B) Chloride
(C) Potassium (D) Magnesium

447. Most abundant element in the Earth crust is __________

(A) Oxygen (B) Silicon
(C) Aluminum (D) Iron

448. First world climate conference held on 12-13, 1979 in __________

(A) Geneva (B) Bonn (C) Brazil (D) USA

449. The first UNFCCC conference of the parties took place from 28th March to 7th April 1995 in __________

(A) Geneva (B) Bonn (C) Brazil (D) Berlin

450. IPCC working group responsible for assessing the mitigation of climate change is __________

(A) WG I (B) WG II (C) WG III (D) WG IV

451. Cartagena protocol call enforce __________

(A) Bioremediation (B) Biomining
(C) Biosafety (D) Biocontrol

452. Resilient time of Glomalin __________

(A) 50 years (B) 500 years
(C) 5000 years (D) 1000 years

453. UN convention to combat Desertification (UNCCD) came in to force in the year __________

(A) 1986 (B) 1996
(C) 2006 (D) 2016

454. Phytoliths are also called as

(A) Plant stone (B) Animal stone
(C) Bacterial stone (D) All

455. CropSyst model was developed by ________

(A) Australia state university
(B) Washington state university
(C) NUS
(D) ICAR

456. __________ can be regarded as pastures of sea

(A) Phytoplankton (B) Zooplankton
(C) Both (D) None

457. Which among the following is not a structural adaptation?

(A) Webbed feet of Duck
(B) Bill on a bird
(C) Fur on a bear
(D) Tanning skin on exposure to sunlight

458. Which among the following is a physiological adaptation?

(A) Webbed feet of Duck
(B) Bill on a bird
(C) Fur on a bear
(D) Tanning skin on exposure to sunlight

459. Chameleon exhibits ________

(A) Physical adaptation (B) Structural adaptation
(C) Physiological adaptation (D) All

460. WOFOST Model is a simulation model for the ________

(A) Quantitative analysis of the growth and production of annual field crops
(B) Qualitative analysis of the growth and production of annual field crops
(C) Both A and B
(D) None of these

461. GWP of N_2O

(A) 34 (B) 265 (C) 1 (D) 65

462. The term La Nina (Spanish word) means

(A) The little girl (B) The little boy
(C) The little man (D) The little women

463. Heat balance of earth is maintained by __________, as it absorbs IR radiation

(A) GHG (B) Atmosphere
(C) Ozone layer (D) Clouds

464. Stratosphere is characterized by
(A) Positive lapse rate (B) Negative lapse rate
(C) Neutral lapse rate (D) None

465. Troposphere is characterized by
(A) Positive lapse rate (B) Negative lapse rate
(C) Neutral lapse rate (D) None

466. GHG based on their contribution to Global warming is in the order of
(A) $CO_2>CH_4>O_3>N_2O$ (B) $CO_2>CH_4>N_2O>O_3$
(C) $CH_4>CO_2>N_2O>O_3$ (D) $O_3>CO_2>N_2O>CH_4$

467. Deadliest skin cancer associated with ozone depletion
(A) Malignant melanoma (B) Arthralgia
(C) Carcinoma (D) Bronchiogenic Carcinoma

468. In the year ________, Nobel peace prize was shared by IPCC and Al Gore for the work on climate change
(A) 2007 (B) 2008
(C) 2009 (D) 2010

469. Second commitment period of Doha amendment to the Kyoto Protocol (2012) starts from
(A) 2013 (B) 2014
(C) 2015 (D) 2016

470. Which among the following is considered as NOT a response to climate change?
(A) Decrease in pollen longevity
(B) Decrease in pollen germination
(C) Decrease in pollen viability
(D) Decrease in seed set

471. Emission of more GHGs in the permafrost region due to carbon mobilizations is considered as
(A) Carbon capture (B) Carbon sink
(C) Microbial prime effect (D) Dung heap effect

472. Oxygenase enzyme activity occurs at ________ condition
(A) Low CO_2, high O_2 (B) Low CO_2, low O_2
(C) High CO_2, high O_2 (D) High CO_2, low O_2

473. Oxygenase enzyme absent in ______ plants
(A) C4 (B) C3
(C) CAM (D) All

474. C4 plants uses alternate enzyme for the first step of carbon fixation is ________

(A) Oxygenase (B) Phosphoenolpyruvate (PEP)
(C) Carboxynase (D) None

475. Convention for substance depletes ozone

(A) Montreal protocol (B) Kyoto protocol
(C) Vienna convention (D) Rio summit

476. Treaty of treaties

(A) Montreal protocol (B) Kyoto protocol
(C) Vienna convention (D) Rio summit

477. Vienna convention on the law of treaties, an international agreement governing treaties between states that was drafted by International Law Commission of the UN was adopted on

(A) May 23, 1969 (B) May 23, 1970
(C) May 23, 1971 (D) May 23, 1972

478. Vienna convention on the law of treaties, an international agreement governing treaties between states that was drafted by International Law Commission of the UN was entered into force on

(A) May 23, 1969 (B) May 23, 1970
(C) January 27, 1971 (D) January 27, 1980

479. Sulphate aerosols connected with

(A) Global warning (B) Global cooling
(C) Sea level rice (D) Melting of glaciers

480. Assessing the status of Himalayan Pheasant is carried out by

(A) PETA (B) BNHS
(C) WWF (D) EPA

481. Which of the three countries have signed a deal in Winnipeg to tap into energy cooperation?

(A) Canada, UK, India (B) Canada, Mexico, USA
(C) Canada, Australia, USA (D) Canada, Australia, India

482. GOSAT developed by

(A) USA (B) UK
(C) Japan (D) India

483. Rubisco found in

(A) C3 (B) C4
(C) CAM (D) All

484. CER denotes
(A) Certified Emission Reduction
(B) Carbon Emission Reduction
(C) Carbon Emission Rate
(D) Certified Emission Rate

485. Documentary showing how farmers fight climate change is
(A) Adaptation Japan (B) Adaptation Bangladesh
(C) Adaptation India (D) Adaptation Kenya

486. If climate becomes warm, Sea level will
(A) Fall (B) Rise
(C) Remain unchanged (D) None

487. Swedish environmental activist who has gained international recognition for promoting view on the crisis of climate change
(A) Greta Tintin Eleonora Ernman Thunberg
(B) Rachael Carson
(C) Al Gore
(D) Keelings

488. Open Top Chambers are used for
(A) Studying the elevated level of CO_2 on plants
(B) Studying the elevated level of CO_2 on animals
(C) Studying the effect of O_3 on plants
(D) All

489. Ground level ozone causes
(A) Respiratory problem (B) Blindness
(C) Cancer (D) CNS

490. Which among the following is a non atmospheric parameter that can be used to study climate change?
(A) Rain (B) Solar radiation
(C) Ice cover (D) Air Temperature

491. GWP of methane over 100 years, according to IPCC
(A) 34 (B) 44 (C) 54 (D) 64

492. Which among the following is related to India's climate change policy?
(A) TNPCB (B) ITCC
(C) NAPCC (D) IPCC

493. Nutrient cycle associated with Bat Guana
(A) N (B) P (C) S (D) C

494. Last IPCC report as on year 2020
(A) AR3 (B) AR4
(C) AR5 (D) AR6

495. IPCC special report on Global warming focus on the temperature increase of
(A) 1.5 °C above preindustrial levels and related global GHG emission pathway
(B) 2.0 °C above preindustrial levels and related global GHG emission pathway
(C) 2.5 °C above preindustrial levels and related global GHG emission pathway
(D) 3.0 °C above preindustrial levels and related global GHG emission pathway

496. Satellite to be launched by Russia for monitoring Arctic climate and environment by the end of December 2020 is
(A) Nimbus 7 (B) Nimbus 2020
(C) Arktika-M (D) Arktika-R

497. Climate change study in polar is important because
(A) Polar ice is depleting at faster rate
(B) Depleting ice releases CH_4
(C) The ice in the polar region reflects incoming solar radiation leads to ocean acidification
(D) All

498. COP 26 scheduled to be held in Glasgow, Scotland during November 2020 was postponed due to the outbreak of
(A) COVID 19 Pandemic (B) Ebola Pandemic
(C) SARS Pandemic (D) Dengue Pandemic

499. About ______ % of species in the world would face the risk of extinction because of climate factors
(A) 16% (B) 20%
(C) 26% (D) 30%

500. International centre for integrated mountain development (ICIMOD) founded in 1983 is located at
(A) Nepal (B) India
(C) Bangladesh (D) Russia

Answers Key

Q. Ans	Q. Ans	Q. Ans	Q. Ans	Q. Ans	Q. Ans	Q. Ans
1. B	2. A	3. B	4. C	5. D	6. C	7. D
8. D	9. D	10. C	11. D	12. B	13. C	14. C
15. A	16. B	17. C	18. A	19. B	20. C	21. A
22. C	23. B	24. B	25. C	26. D	27. A	28. A
29. C	30. B	31. D	32. D	33. A	34. C	35. B
36. D	37. B	38. B	39. D	40. A	41. C	42. D
43. B	44. A	45. A	46. A	47. A	48. C	49. A
50. D	51. B	52. A	53. C	54. A	55. B	56. B
57. B	58. A	59. C	60. D	61. D	62. A	63. C
64. A	65. D	66. B	67. B	68. A	69. D	70. B
71. C	72. A	73. D	74. B	75. D	76. A	77. C
78. C	79. C	80. A	81. A	82. B	83. A	84. B
85. C	86. B	87. B	88. A	89. D	90. D	91. A
92. A	93. C	94. A	95. A	96. B	97. B	98. C
99. C	100. B	101. D	102. C	103. B	104. D	105. A
106. C	107. A	108. B	109. B	110. A	111. C	112. C
113. B	114. B	115. C	116. B	117. C	118. B	119. A
120. B	121. C	122. D	123. A	124. A	125. D	126. A
127. B	128. B	129. A	130. A	131. B	132. B	133. B
134. B	135. B	136. B	137. B	138. C	139. C	140. C
141. A	142. A	143. D	144. A	145. D	146. A	147. D
148. A	149. D	150. B	151. C	152. C	153. A	154. B
155. A	156. C	157. C	158. A	159. A	160. B	161. D
162. A	163. A	164. A	165. D	166. A	167. D	168. D
169. B	170. C	171. A	172. D	173. A	174. C	175. D
176. A	177. A	178. D	179. A	180. B	181. A	182. B
183. A	184. B	185. C	186. B	187. A	188. A	189. D
190. B	191. C	192. A	193. C	194. C	195. A	196. B
197. A	198. C	199. B	200. D	201. A	202. C	203. C
204. C	205. C	206. B	207. C	208. D	209. C	210. C
211. B	212. D	213. B	214. A	215. A	216. B	217. A
218. B	219. B	220. B	221. D	222. C	223. C	224. C
225. D	226. D	227. D	228. C	229. C	230. D	231. C
232. D	233. D	234. C	235. B	236. A	237. B	238. D
239. D	240. A	241. C	242. B	243. D	244. D	245. B
246. A	247. A	248. B	249. B	250. A	251. D	252. C

Q.	Ans	Q.	Ans	Q.	Ans	Q.	Ans	Q.	Ans	Q.	Ans	Q.	Ans
253.	A	254.	B	255.	B	256.	C	257.	D	258.	D	259.	C
260.	D	261.	B	262.	B	263.	A	264.	B	265.	C	266.	A
267.	C	268.	C	269.	B	270.	D	271.	A	272.	A	273.	D
274.	A	275.	B	276.	A	277.	C	278.	A	279.	B	280.	A
281.	A	282.	D	283.	C	284.	A	285.	C	286.	B	287.	C
288.	D	289.	D	290.	C	291.	A	292.	A	293.	B	294.	A
295.	A	296.	D	297.	B	298.	C	299.	C	300.	B	301.	A
302.	C	303.	B	304.	B	305.	B	306.	C	307.	C	308.	B
309.	D	310.	A	311.	A	312.	D	313.	D	314.	A	315.	A
316.	B	317.	C	318.	D	319.	A	320.	A	321.	B	322.	B
323.	D	324.	D	325.	D	326.	B	327.	B	328.	C	329.	A
330.	C	331.	C	332.	B	333.	A	334.	A	335.	A	336.	D
337.	A	338.	B	339.	C	340.	C	341.	D	342.	A	343.	B
344.	A	345.	D	346.	B	347.	C	348.	C	349.	B	350.	A
351.	A	352.	C	353.	A	354.	B	355.	C	356.	A	357.	B
358.	D	359.	B	360.	A	361.	A	362.	A	363.	B	364.	C
365.	D	366.	A	367.	B	368.	C	369.	B	370.	A	371.	B
372.	A	373.	B	374.	D	375.	C	376.	B	377.	C	378.	A
379.	D	380.	D	381.	B	382.	B	383.	A	384.	A	385.	B
386.	A	387.	C	388.	A	389.	D	390.	A	391.	A	392.	C
393.	D	394.	B	395.	C	396.	D	397.	A	398.	A	399.	C
400.	B	401.	D	402.	A	403.	C	404.	C	405.	D	406.	A
407.	A	408.	B	409.	D	410.	A	411.	D	412.	C	413.	A
414.	B	415.	B	416.	A	417.	A	418.	C	419.	B	420.	B
421.	D	422.	B	423.	B	424.	B	425.	B	426.	B	427.	B
428.	B	429.	B	430.	B	431.	C	432.	D	433.	B	434.	C
435.	A	436.	A	437.	A	438.	C	439.	C	440.	C	441.	A
442.	B	443.	B	444.	C	445.	D	446.	B	447.	A	448.	A
449.	D	450.	C	451.	C	452.	A	453.	B	454.	A	455.	B
456.	A	457.	D	458.	D	459.	A	460.	A	461.	B	462.	A
463.	A	464.	B	465.	A	466.	A	467.	A	468.	A	469.	A
470.	A	471.	C	472.	A	473.	A	474.	B	475.	A	476.	C
477.	A	478.	D	479.	B	480.	B	481.	B	482.	C	483.	A
484.	A	485.	B	486.	B	487.	A	488.	A	489.	A	490.	C
491.	A	492.	C	493.	B	494.	C	495.	A	496.	C	497.	D
498.	A	499.	A	500.	A								

CHAPTER 14

IMPORTANT ENVIRONMENTAL DAYS OBSERVED WORLD WIDE

1. **February 2 (World Wetland Day)**

   On this day, in 1971, the Ramsar Convention on Wetlands of International Importance was signed. Wetlands are a very important part of our biodiversity and it is essential to see that they are well protected.

2. **February 28 (National Science Day)**

   It is necessary to highlight the role of science in the protection of the environment. This day should be taken as a platform to put forward the message.

3. **March 21 (World Forestry Day)**

   Activities such as the planting of trees and highlighting, the urgency to increase the green cover.

4. **March 22 (World Water Day)**

   The decision to celebrate this day has been taken recently as drinking water sources are fast depleting. The world must wake up to the problem and begin conserving it.

5. **March 23 (World Meteorological Day)**

   Everyone has to be reminded that weather is an integral part of the environment.

6. **April 7 (World Health Day)**

   The World Health Organization (WHO) was constituted on this day in 1948. In the changing environment around us health is becoming an important issue.

7. **April 18 (World Heritage Day)**

   Environment includes not just the natural surroundings but also the manmade ones.

8. **April 22 (Earth Day)**

   In 1970 a group of people in the United States of America got together to draw the attention of the world to the problems being caused to the earth due to modernization. Since then this day has been celebrated all over the world as Earth Day.

9. **May 31 (Anti Tobacco Day)**

   The world is now aware of the problems faced by not only the smokers but also the people who inhale the smoke. You can take up an anti-smoking campaign in your family or the neighborhood.

10. **June 5 (World Environment Day)**

    On this day, in 1972, the Stockholm Conference on Human Environment was held in Sweden. There was a large gathering from all over the world and people expressed their concerns for the increasing environmental problems.

11. **July 11 (World Population Day)**

    Population has to be given special attention, as it is an ever-increasing problem especially in India.

12. **September 16 (World Ozone Day)**

    The United Nations declared this day as the International Day for the Preservation of the Ozone Layer. It is the day the Montreal Protocol was signed.

13. **September 28 (Green Consumer Day)**

    The problems of consumerism and its impact on the environment is an area of major concern in today's world. Awareness building on the importance of recycling- reusing-reducing should be taken up seriously.

14. **October 3 (World Habitat Day)**

    The earth is the habitat of not only human beings but also all living creatures. Increasing human activities is threatening the habitat of other living things.

15. **October 1-7 (World Wildlife Week)**

    Celebrate this week by building awareness on the importance of preservation of our wildlife.

16. **October 4 (World Animal Welfare Day)**

    The welfare of animals has to be looked into and given due importance.

17. October 13 (International Day for Natural Disaster Reduction)

Due to a change in the environment there has been an increase in the number of natural disasters. Efforts have to be taken to reduce these disasters.

18. October 16 (World Food Day)

19. November 14 (Children's Day in India)

Children can work together for a better tomorrow by improving the environment around them.

20. December 2 - 3 (Bhopal Tragedy Day)

Mark this occasion by taking a pledge to put in your best efforts to prevent such a tragedy from occurring again.

◈

CHAPTER 15

WORLD ENVIRONMENT DAY THEME AND HOST CITIES

Year	Theme	Host city
1974	Only one Earth during Expo '74	Spokane, United States
1975	Human Settlements	Dhaka, Bangladesh
1976	Water: Vital Resource for Life	Ontario, Canada
1977	Ozone Layer Environmental Concern; Lands Loss and Soil Degradation	Sylhet, Bangladesh
1978	Development Without Destruction	Sylhet, Bangladesh
1979	Only One Future for Our Children – Development Without Destruction	Sylhet, Bangladesh
1980	A New Challenge for the New Decade: Development Without Destruction	Sylhet, Bangladesh
1981	Ground Water; Toxic Chemicals in Human Food Chains	Sylhet, Bangladesh
1982	Ten Years After Stockholm (Renewal of Environmental Concerns)	Dhaka, Bangladesh
1983	Managing and Disposing Hazardous Waste: Acid Rain and Energy	Sylhet, Bangladesh
1984	Desertification	Rajshahi, Bangladesh
1985	Youth: Population and the Environment	Islamabad, Pakistan
1986	A Tree for Peace	Ontario, Canada

Year	Theme	Host city
1987	Environment and Shelter: More Than A Roof	Nairobi, Kenya
1988	When People Put the Environment First, Development Will Last	Bangkok, Thailand
1989	Global Warming; Global Warning	Brussels, Belgium
1990	Children and the Environment	Mexico City, Mexico
1991	Climate Change. Need for Global Partnership	Stockholm, Sweden
1992	Only One Earth, Care and Share	Rio de Janeiro, Brazil
1993	Poverty and the Environment – Breaking the Vicious Circle	Beijing, People's Republic of China
1994	One Earth One Family	London, United Kingdom
1995	We the Peoples: United for the Global Environment	Pretoria, South Africa
1996	Our Earth, Our Habitat, Our Home	Istanbul, Turkey
1997	For Life on Earth	Seoul, Republic of Korea
1998	For Life on Earth – Save Our Seas	Moscow, Russian Federation
1999	Our Earth – Our Future – Just Save It	Tokyo, Japan
2000	The Environment Millennium – Time to Act	Adelaide, Australia
2001	Connect with the World Wide Web of Life	Torino, Italy and Havana, Cuba
2002	Give Earth a Chance	Shenzhen, People's Republic of China
2003	Water – Two Billion People are Dying for It!	Beirut, Lebanon

Year	Theme	Host city
2004	Wanted! Seas and Oceans – Dead or Alive?	Barcelona, Spain
2005	Green Cities – Plan for the Planet!	San Francisco, United States
2006	Deserts and Desertification – Don't Desert Dry lands	Algiers, Algeria
2007	Melting Ice – a Hot Topic?	London, England
2008	Kick The Habit – Towards A Low Carbon Economy	Wellington, New Zealand
2009	Your Planet Needs You – Unite to Combat Climate Change	Mexico City, Mexico
2010	Many Species. One Planet. One Future	Rangpur, Bangladesh
2011	Forests: Nature at your Service	New Delhi, India
2012	Green Economy: Does it include you?	Brasilia, Brazil
2013	Think. Eat. Save. Reduce Your Food print	Ulaanbaatar, Mongolia
2014	Raise your voice, not the sea level	Bridgetown, Barbados
2015	Seven Billion Dreams. One Planet. Consume with Care.	Rome, Italy
2016	Zero Tolerance for the Illegal Wildlife trade	Luanda, Angola
2017	Connecting People to Nature – in the city and on the land, from the poles to the equator	Ottawa, Canada
2018	Beat Plastic Pollution	New Delhi, India
2019	Beat Air Pollution	China
2020	Biodiversity : Time for Nature	Colombia in partnership with Germany

CHAPTER 16

WORLD WATER DAY THEME

World Water Day is an annual UN observance day (22 March) that highlights the importance of freshwater. The day is used to advocate for the sustainable management of freshwater resources. The theme of each day focuses on topics relevant to clean water, sanitation and hygiene (WASH), which is in line with the targets of Sustainable Development Goal 6. The UN World Water Development Report (WWDR) is released each year around World Water Day. UN-Water is the convener for World Water Day and selects the theme for each year in consultation with UN organizations that share an interest in that year's focus. World Water Day is celebrated around the world with a variety of events. These can be theatrical, musical or lobbying in nature. The day can also include campaigns to raise money for water projects. The first World Water Day, designated by the United Nations, was in 1993.

- 2020 : Water and Climate Change
- 2019 : Leaving No One Behind
- 2018 : Nature for Water
- 2017 : Why Waste Water
- 2016 : Better Water, Better Jobs
- 2015 : Water and Sustainable Development
- 2014 : Water and Energy
- 2013 : International Year of Cooperation
- 2012 : Water and Food Security
- 2011 : Water for cities: responding to the urban challenge
- 2010 : Clean Water for a Healthy World

- 2009 : Trans Waters
- 2008 : Sanitation. 2008 was also the International Year of Sanitation
- 2007 : Coping With Water Scarcity
- 2006 : Water and Culture
- 2005 : Water for Life
- 2004 : Water and Disasters
- 2003 : Water for Future
- 2002 : Water for Development
- 2001 : Water for Health
- 2000 : Water for the 21st century
- 1999 : Everyone Lives Downstream
- 1998 : Groundwater - The Invisible Resource.
- 1997 : The World's Water: Is there enough?
- 1996 : Water for Thirsty Cities
- 1995 : Women and Water
- 1994 : Caring for our Water Resources is Everybody's Business.

◈

CHAPTER 17

AIR QUALITY STANDARDS & AIR QUALITY INDEX (AQI)

AMBIENT AIR QUALITY STANDARDS IN INDIA

Pollutant	Time Weighted Average	Concentration in Ambient Air	
		Industrial, Residential, Rural and Other Areas	Ecologically Sensitive Area (notified by Central Government)
Sulphur Dioxide (SO_2), $\mu g/m^3$	Annual* 24 hours**	50 80	20 80
Nitrogen Dioxide (NO_2), $\mu g/m^3$	Annual* 24 hours**	40 80	30 80
Particulate Matter (size less than 10 µm) or PM_{10} $\mu g/m^3$	Annual* 24 hours**	60 100	60 100
Particulate Matter (size less than 2.5 µm) or $PM_{2.5}$ $\mu g/m^3$	Annual* 24 hours**	40 60	40 60
Ozone (O_3), $\mu g/m^3$	8 hours* 1 hour**	100 180	100 180
Lead (Pb), $\mu g/m^3$	Annual* 24 hours**	0.50 1.00	0.50 1.00
Carbon Monoxide (CO), mg/m^3	8 hours* 1 hour**	02 04	02 04
Ammonia (NH_3), $\mu g/m^3$	Annual* 24 hours**	100 400	100 400

Pollutant	Time Weighted Average	Concentration in Ambient Air	
		Industrial, Residential, Rural and Other Areas	Ecologically Sensitive Area (notified by Central Government)
Benzene (C_6H_6), µg/m^3	Annual*	5	5
Benzo(a)Pyrene (BaP)-particulate phase only, ng/m^3	Annual*	1	1
Arsenic(As), ng/m^3	Annual*	6	60
Nickel (Ni), ng/m^3	Annual*	20	20

* Annual arithmetic mean of minimum 104 measurements in a year at a particular site taken twice a week 24 hourly at uniform intervals.

** 24 hourly or 8 hourly or 1 hourly monitored values, as applicable, shall be complied with 98% of the time, they may exceed the limits but not on two consecutive days of monitoring.

Source: National Ambient Air Quality Standards, Central Pollution Control Board Notification in the Gazette of India, Extraordinary, New Delhi, 18th November, 2009

AIR QUALITY RANGE

AQI Category, Pollutants and Health Breakpoints

AQI Category (Range)	PM_{10} (24 hr)	$PM_{2.5}$ (24 hr)	NO_2 (24 hr)	O_3 (8hr)	CO (8hr)	SO_2 (24 hr)	NH_3 (24 hr)	Pb (24 hr)
Good (0–50)	0–50	0–30	0–40	0–50	0–1.0	0–40	0–200	0–0.5
Satisfactory (51–100)	51–100	31–60	41–80	51–100	1.1–2.0	41–80	201–400	0.5–1.0
Moderately polluted (101–200)	101–250	61–90	81–180	101–168	2.1–10	81–380	401–800	1.1–2.0

AQI Category (Range)	PM_{10} (24 hr)	$PM_{2.5}$ (24 hr)	NO_2 (24 hr)	O_3 (8hr)	CO (8hr)	SO_2 (24 hr)	NH_3 (24 hr)	Pb (24 hr)
Poor (201–300)	251–350	91–120	181–280	169–208	10–17	381–800	801–1200	2.1–3.0
Very poor (301–400)	351–430	121–250	281–400	209–748	17–34	801–1600	1200–1800	3.1–3.5
Severe (401–500)	430+	250+	400+	748+	34+	1600+	1800+	3.5+

AIR QUALITY INDEX (AQI)

AQI	Associated Health Impacts
Good (0–50)	Minimal impact
Satisfactory (51–100)	May cause minor breathing discomfort to sensitive people.
Moderately polluted (101–200)	May cause breathing discomfort to people with lung disease such as asthma, and discomfort to people with heart disease, children and older adults.
Poor (201–300)	May cause breathing discomfort to people on prolonged exposure, and discomfort to people with heart disease.
Very poor (301–400)	May cause respiratory illness to the people on prolonged exposure. Effect may be more pronounced in people with lung and heart diseases.
Severe (401–500)	May cause respiratory impact even on healthy people, and serious health impacts on people with lung/heart disease. The health impacts may be experienced even during light physical activity.

CHAPTER 18

WATER QUALITY STANDARDS

Water quality standards are benchmarks established to assess whether the quality of rivers and lakes is adequate for fish and other aquatic life, recreation, drinking, agriculture, industry and other uses.

International Standard for Drinking Water Quality (WHO)

S. No.	Parameter	Units	Maximum permissible limit
1	pH	–	6.5-9.2
2	Ar	mg L^{-1}	0.05
3	NH_3	mg L^{-1}	0.5
4	BOD	mg L^{-1}	6.0
5	Ca	mg L^{-1}	100
6	Cd	mg L^{-1}	0.01
7	Cr	mg L^{-1}	0.05
8	Cu	mg L^{-1}	1.5
9	Cl	mg L^{-1}	500
10	Cyanide	mg L^{-1}	0.05
11	COD	mg L^{-1}	10
12	Fe	mg L^{-1}	1.0
13	Pb	mg L^{-1}	0.1
14	Mg	mg L^{-1}	150
15	Mn	mg L^{-1}	0.5
16	Hg	mg L^{-1}	0.001
17	Nitrate and Nitrite	mg L^{-1}	45
18	Phenol	mg L^{-1}	0.002
19	PAH	mg L^{-1}	0.2

S. No.	Parameter	Units	Maximum permissible limit
20	Pesticides	mg L^{-1}	–
21	*E.coli*	–	10/100ml
22	Total hardness	mg L^{-1}	500
23	TDS	mg L^{-1}	500

IRRIGATION WATER QUALITY STANDARDS (FOR DISCHARGE)

S. No.	Parameter	Inland surface water
1.	Colour and odour	Colorless and odorless
2.	Suspended Solids, mg/l, Max	100
3.	Particle size of suspended solids	Shall pass 850 micron IS Sieve
4.	Dissolved solids (inorganic), mg/l	2100
5.	pH value	5.5 to 9.0
6.	Temperature °C, Max	Shall not exceed 40 in any section of the stream within 15 meters down stream from the effluent outlet
7.	Oil and grease, mg/l, max	10
8.	Total residual chlorine, mg/l, Max.	1.0
9.	Ammonical nitrogen (as N), mg/l, Max.	50
10.	Total Kjeldahl nitrogen (as N), mg/l, Max.	100
11.	Free Ammonia (as NH_3), mg/l, Max.	5.0
12.	Biochemical Oxygen Demand (5 days at 20°C) Max.	30
13.	Chemical Oxygen Demand, mg/l, Max.	250

S. No.	Parameter	Inland surface water
14.	Arsenic (as As), mg/l, Max.	0.2
15.	Mercury (as Hg), mg/l, Max.	0.01
16.	Lead (as Pb), mg/l, Max.	0.1
17.	Cadmium (as Cd), mg/l, Max.	2.0
18.	Hexavalent chromium (as Cr+6) mg/l, Max.	0.1
19.	Total chromium as (Cr), mg/l, Max.	2.0
20.	Copper (as Cu), mg/l, Max.	3.0
21.	Zinc (as Zn), mg/l, Max.	5.0
22.	Selenium (as Se), mg/l, Max.	0.05
23.	Nickel (as Ni), mg/l, Max.	3.0
24.	Boron (as B), mg/l, Max.	2.0
25.	Percent Sodium, Max.	–
26.	Residual sodium carbonate, mg/l, Max.	–
27.	Cyanide (as CN), mg/l, Max.	0.2
28.	Chloride (as Cl), mg/l, Max.	1000
29.	Fluoride (as F), mg/l, Max.	2.0
30.	Dissolved Phosphates (as P), mg/l, Max.	5.0
31.	Sulphate (as SO_4), mg/l, Max.	1000
32.	Sulphide (as S), mg/l, Max.	2.0
33.	Pesticides	Absent
34.	Phenolic compounds (as C_6H_5OH), mg/l, Max.	1.0
35.	Radioactive materials	
	(a) Alpha emitters MC/ml, Max.	10^{-7}
	(b) Beta emitters uc/ml, Max.	10^{-6}

◈

CHAPTER 19

NOISE QUALITY STANDARDS

Place	Limits dB	
	Day time	Night time
Industrial area	75	70
Commercial area	65	55
Residential area	55	45
Silent zone	50	40

Noise Source	Intensity
Breathing	10 dB
Soft whisper	20 dB
Library	30 dB
Low volume radio (Walkman)	35 dB
Normal conversation	35 to 60 dB
Telephone	60 dB
Alarm clock	70 dB
Traffic	70 to 90 dB
Lions roar (at 12 feet)	105 dB
Thunder, Jet, Train whistle (at 50 feet)	110 dB
Air craft Jet take off (at 100 feet)	120 dB
Siren	150 dB
Space rocket	170 dB

CHAPTER 20

IMPORTANT CLIMATE FACTS

SHARED SOCIOECONOMIC PATHWAYS

Shared Socioeconomic Pathways (SSPs) are scenarios of projected socioeconomic global changes up to 2100. They are used to derive greenhouse gas emissions scenarios with different climate policies.

The scenarios are:

- SSP1 : Sustainability (Taking the Green Road)
- SSP2 : Middle of the Road
- SSP3 : Regional Rivalry (A Rocky Road)
- SSP4 : Inequality (A Road divided)
- SSP5 : Fossil-fueled Development (Taking the Highway)

They will be used to help produce the IPCC Sixth Assessment Report on global warming, due in 2021.

SSP1 : Sustainability (Taking the Green Road)

The world shifts gradually, but pervasively, toward a more sustainable path, emphasizing more inclusive development that respects perceived environmental boundaries. Management of the global commons slowly improves, educational and health investments accelerate the demographic transition, and the emphasis on economic growth shifts toward a broader emphasis on human well-being. Driven by an increasing commitment to achieving development goals, inequality is reduced both across and within countries. Consumption is oriented toward low material growth and lower resource and energy intensity.

SSP2 : Middle of the road

The world follows a path in which social, economic, and technological trends do not shift markedly from historical patterns. Development and income growth proceeds unevenly, with some countries making relatively good progress while others fall short of expectations. Global and national institutions work toward but make slow progress in achieving sustainable development goals. Environmental systems experience degradation, although there are some improvements and overall the intensity of resource and energy use declines. Global population growth is moderate and levels off in the second half of the century. Income inequality persists or improves only slowly and challenges to reducing vulnerability to societal and environmental changes remain.

SSP3 : Regional rivalry (A Rocky Road)

A resurgent nationalism, concerns about competitiveness and security, and regional conflicts push countries to increasingly focus on domestic or, at most, regional issues. Policies shift over time to become increasingly oriented toward national and regional security issues. Countries focus on achieving energy and food security goals within their own regions at the expense of broader-based development. Investments in education and technological development decline. Economic development is slow, consumption is material-intensive, and inequalities persist or worsen over time. Population growth is low in industrialized and high in developing countries. A low international priority for addressing environmental concerns leads to strong environmental degradation in some regions.

SSP4 : Inequality (A Road Divided)

Highly unequal investments in human capital, combined with increasing disparities in economic opportunity and political power, lead to increasing inequalities and stratification both across and within countries. Over time, a gap widens between an internationally-connected society that contributes to knowledge- and capital-intensive sectors of the global economy, and a fragmented collection of lower-income, poorly educated societies that work in a labor intensive, low-tech economy. Social cohesion degrades and conflict and unrest become increasingly common. Technology development is high in the high-tech economy and sectors. The globally connected energy sector diversifies, with investments in both carbon-intensive fuels like coal and unconventional oil, but also low-carbon energy sources.

Environmental policies focus on local issues around middle and high income areas.

SSP5 : Fossil-Fueled Development (Taking the Highway)

This world places increasing faith in competitive markets, innovation and participatory societies to produce rapid technological progress and development of human capital as the path to sustainable development. Global markets are increasingly integrated. There are also strong investments in health, education, and institutions to enhance human and social capital. At the same time, the push for economic and social development is coupled with the exploitation of abundant fossil fuel resources and the adoption of resource and energy intensive lifestyles around the world. All these factors lead to rapid growth of the global economy, while global population peaks and declines in the 21st century. Local environmental problems like air pollution are successfully managed. There is faith in the ability to effectively manage social and ecological systems, including by geo-engineering if necessary.

SPECIAL REPORT ON EMISSIONS SCENARIOS

The Special Report on Emissions Scenarios (SRES) is a report by the Intergovernmental Panel on Climate Change (IPCC) that was published in 2000. The greenhouse gas emissions scenarios described in the Report have been used to make projections of possible future climate change. The SRES scenarios, as they are often called, were used in the IPCC Third Assessment Report (TAR), published in 2001, and in the IPCC Fourth Assessment Report (AR4), published in 2007.

The SRES scenarios were designed to improve upon some aspects of the IS92 scenarios, which had been used in the earlier IPCC Second Assessment Report of 1995. The SRES scenarios are "baseline" (or "reference") scenarios, which means that they do not take into account any current or future measures to limit greenhouse gas (GHG) emissions (e.g., the Kyoto Protocol to the United Nations Framework Convention on Climate Change). Emissions projections of the SRES scenarios are broadly comparable in range to the baseline emissions scenarios that have been developed by the scientific community. The SRES scenarios, however, do not encompass the full range of possible futures: emissions may change less than the scenarios imply, or they could change more.

Scenario families contain individual scenarios with common themes. The six families of scenarios discussed in the IPCC's Third Assessment Report (TAR) and Fourth Assessment Report (AR4) are A1FI, A1B, A1T, A2, B1, and B2. The IPCC did not state that any of the SRES scenarios were more likely to occur than others, therefore none of the SRES scenarios represent a "best guess" of future emissions. Scenario descriptions are based on those in AR4, which are identical to those in TAR.

A1

The A1 scenarios are of a more integrated world. The A1 family of scenarios is characterized by:

- Rapid economic growth.
- A global population that reaches 9 billion in 2050 and then gradually declines.
- The quick spread of new and efficient technologies.
- A convergent world - income and way of life converge between regions. Extensive social and cultural interactions worldwide.

There are subsets to the A1 family based on their technological emphasis:

- A1FI - An emphasis on fossil-fuels (Fossil Intensive).
- A1B - A balanced emphasis on all energy sources.
- A1T - Emphasis on non-fossil energy sources.

A2

The A2 scenarios are of a more divided world. The A2 family of scenarios is characterized by:

- A world of independently operating, self-reliant nations.
- Continuously increasing population.
- Regionally oriented economic development.
- Low emissions

B1

The B1 scenarios are of a world more integrated, and more ecologically friendly. The B1 scenarios are characterized by:

- Rapid economic growth as in A1, but with rapid changes towards a service and information economy.
- Population rising to 9 billion in 2050 and then declining as in A1.
- Reductions in material intensity and the introduction of clean and resource efficient technologies.
- An emphasis on global solutions to economic, social and environmental stability.

B2

The B2 scenarios are of a world more divided, but more ecologically friendly. The B2 scenarios are characterized by:

- Continuously increasing population, but at a slower rate than in A2.
- Emphasis on local rather than global solutions to economic, social and environmental stability.
- Intermediate levels of economic development.
- Less rapid and more fragmented technological change than in A1 and B1.

SUSTAINABLE DEVELOPMENT GOALS (SDGS) TO TRANSFORM OUR WORLD

GOAL 1 : No Poverty

GOAL 2 : Zero Hunger

GOAL 3 : Good Health and Well-being

GOAL 4 : Quality Education

GOAL 5 : Gender Equality

GOAL 6 : Clean Water and Sanitation

GOAL 7 : Affordable and Clean Energy

GOAL 8 : Decent Work and Economic Growth

GOAL 9 : Industry, Innovation and Infrastructure

GOAL 10 : Reduced Inequality

GOAL 11 : Sustainable Cities and Communities

GOAL 12	:	Responsible Consumption and Production
GOAL 13	:	Climate Action
GOAL 14	:	Life below Water
GOAL 15	:	Life on Land
GOAL 16	:	Peace and Justice Strong Institutions
GOAL 17	:	Partnerships to achieve the Goal

UNITED NATIONS CLIMATE CHANGE CONFERENCE (CONFERENCE OF THE PARTIES - COP)

1995	:	COP 1,	Berlin, Germany
1996	:	COP 2,	Geneva, Switzerland
1997	:	COP 3,	Kyoto, Japan
1998	:	COP 4,	Buenos Aires, Argentina
1999	:	COP 5,	Bonn, Germany
2000	:	COP 6,	The Hague, Netherlands
2001	:	COP 6	Negotiations, Bonn, Germany
2001	:	COP 7,	Marrakech, Morocco
2002	:	COP 8,	New Delhi, India
2003	:	COP 9,	Milan, Italy
2004	:	COP 10,	Buenos Aires, Argentina
2005	:	COP 11,	Montreal, Canada
2006	:	COP 12,	Nairobi, Kenya
2007	:	COP 13,	Bali, Indonesia
2008	:	COP 14,	Poznań, Poland
2009	:	COP 15,	Copenhagen, Denmark
2010	:	COP 16,	Cancún, Mexico
2011	:	COP 17,	Durban, South Africa
2012	:	COP 18,	Doha, Qatar
2013	:	COP 19,	Warsaw, Poland
2014	:	COP 20,	Lima, Peru
2015	:	COP 21,	Paris, France
2016	:	COP 22,	Marrakech, Morocco
2017	:	COP 23,	Bonn, Germany
2018	:	COP 24,	Katowice, Poland
2019	:	COP 25,	Madrid, Spain

CHAPTER 21

SELF ASSESSMENT TEST

TEST 1 : CLIMATE CHANGE

Choose the correct answer

1. **The heat of sun is not stronger in the area located**
   (A) near the North Pole (B) near the equator
   (C) far from equator (D) near the South Pole

2. **The fifth layer of the Earth's atmosphere is**
   (A) exosphere (B) mesosphere
   (C) thermosphere (D) troposphere

3. **The data collected regarding weather includes**
   (A) precipitation and wind (B) rainfall
   (C) temperature (D) all of above

4. **The places where the temperature of air is above freezing point of the water, the precipitation will be in the form of**
   (A) hailstorms (B) thunders
   (C) rain (D) lightning

5. **The places where the temperature of air is below freezing point of the water, the precipitation will be in the form of**
   (A) snow (B) lightning
   (C) hailstorms (D) thunders

6. **The tropical climates does not includes**
   (A) polar climate (B) monsoon
   (C) savannah (D) equatorial

7. **The shapeless clouds that brings rain for the longer period of time are classified as**
   (A) Stevenson clouds
   (B) cumulus clouds
   (C) stratus clouds
   (D) cirrus clouds

8. The unit of measurement to measure rain is
   (A) degree Fahrenheit
   (B) degree Celsius
   (C) millimeter or centimeter
   (D) percentage

9. As compared to cold air, the warm air is
   (A) unsaturated (B) lighter
   (C) heavier (D) saturated

10. The third layer of the Earth's atmosphere is
    (A) troposphere (B) stratosphere
    (C) mesosphere (D) thermosphere

Match the following

11.	Kyotoprotocol	-	(A) Little boy
12.	Weather	-	(B) Little girl
13.	Climate	-	(C) CO_2
14.	ElNino	-	(D) 1997
15.	LaNino	-	(E) Short term
16.	Cyclone	-	(F) Long term
17.	Anticyclone	-	(G) Aerosols
18.	Rain bearing clouds	-	(H) Cumulo Nimbus
19.	Global warming	-	(I) Low pressure system
20.	Global dimming	-	(J) High pressure system

TEST 1 - KEYS

Q. Ans	Q. Ans	Q. Ans	Q. Ans	Q. Ans	Q. Ans	Q. Ans
1. B	2. A	3. D	4. C	5. A	6. A	7. C
8. C	9. B	10. C	11. D	12. E	13. F	14. A
15. B	16. I	17. J	18. H	19. C	20. G	

TEST 2 : CLIMATE CHANGE

Choose the correct answer

1. The box in which thermometers are kept to save them from direct sunlight is called
   (A) Stevenson Screen (B) Cirrus Screen
   (C) Cumulus Screen (D) Stratus Screen

2. **The human activities that causes climate change on Earth includes**
   (A) burning of forests (B) agricultural activities
   (C) use of aerosol cans (D) all of above
3. **Number of institutions in the Indian Network for Climate Change Assessment (INCCA):**
   (A) 79 (B) 100 (C) 120 (D) 127
4. **The 'thickness' of Stratospheric Ozone layer is measured in**
   (A) Sieverts (B) Dobson units
   (C) Melson units (D) Beaufort Scale
5. **The ranking of countries as the highest per capita emitters of Carbon Dioxide in the world, takes into account their annual CO_2 emissions from**
   1. Burning of fossil fuels
   2. Cement Manufacture
   3. Land Use such as deforestation
   (A) Statement 1 (B) Statement 1 and 2
   (C) Statement 1 and 3 (D) Statement 1, 2 and 3
6. **When was India's first National Action Plan on Climate Change (NAPCC) launched?**
   (A) 2010 (B) 2008
   (C) 2001 (D) 2006
7. **Which of the following is not the mission of India's National Action Plan on Climate Change?**
   (A) National solar mission
   (B) National water mission
   (C) National Mission for a Green India
   (D) National Mission on Sustainable Development
8. **Which Ministry has undertaken the National Mission for a "Green India?**
   (A) Ministry of Rural Development
   (B) Ministry of Environment & Forest
   (C) Ministry of Earth Affairs
   (D) None of these
9. **The United Nations Framework Convention on Climate Change (UNFCCC or FCCC) is held in:**
   (A) 1992 (B) 1995
   (C) 1989 (D) 1990

10. **The Kyoto Protocol, signed in**

(A) Japan in 1997 (B) U.S in 1998

(C) China in 1997 (D) India in 1998

Match the following

11.	GHG	-	(A) 1997
12.	Aerosol	-	(B) Global warming
13.	Montreal protocol	-	(C) 1992
14.	Kyto protocol	-	(D) Global dimming
15.	Earth Summit reviewed	-	(E) Publishing peer findings on climate change in India
16.	Stratosphere	-	(F) UNFCCC
17.	Extra tropical cyclone	-	(G) Smaller mesoscale
18.	Dust devil	-	(H) Largest Scale
19.	NATCOM	-	(I) Ozone
20.	INCCA	-	(J) 1987

TEST 2 - KEYS

Q. Ans	Q. Ans	Q. Ans	Q. Ans	Q. Ans	Q. Ans	Q. Ans
1. A	2. D	3. D	4. B	5. B	6. B	7. D
8. B	9. A	10. A	11. B	12. D	13. J	14. A
15. C	16. I	17. G	18. H	19. F	20. E	

TEST 3 : CLIMATE CHANGE

Choose the correct answer

1. **Climate system that responds more slowly to climate forcing are**

(A) Ocean and ice caps (B) Mountain

(C) Buildings (D) All

2. **The eruption of Mount Pinatubo, the second largest terrestrial eruption of the 20th century, affected the climate substantially, subsequently decreased the global temperatures, happened in the year ________**

(A) 1971 (B) 1981

(C) 1991 (D) 2001

3. **Dendroclimatology is the analysis of ________ to determine past climate variations**
   (A) Tree pollen (B) Fossil palynomorph
   (C) tree ring growth patterns (D) ice core

4. **The Intergovernmental Panel on Climate Change (IPCC) was established by**
   (A) World Meteorological Organization (WMO) and UNEP
   (B) World Health Organization (WHO)
   (C) Environmental Protection Agency (EPA)
   (D) Earth watch and UNEP

5. **Trees play a key role in maintaining soil fertility because**
   (A) Their leaves rot down and fertilize the soil
   (B) They capture the atmospheric Nitrogen
   (C) They live for 100 years
   (D) They absorbs more water

TEST 3 - KEYS

Q. Ans	Q. Ans	Q. Ans	Q. Ans	Q. Ans
1. A	2. C	3. C	4. A	5. A

TEST 4 : POLLUTION

1. **Plastics with thickness of ______ cannot be used for recycling.**
   (A) <30 microns (B) <50microns
   (C) <20 Microns (D) <60 microns

2. **When was the Water (Prevention and control pollution) Act passed?**
   (A) 1972 (B) 1974
   (C) 1986 (D) 1982

3. **Which part of the Indian constitution states the following – "The state shall endeavor to protect and improve the environment and to safeguard the forests and wildlife of the country"?**
   (A) Article 42 A (B) Article 51 A
   (C) Article 56 A (D) Article 48 A

4. **The head quarters of UNEP is in ________**
   (A) Nairobi (B) Brussels
   (C) New York (D) London

5. If a person applies for consent to discharge industrial effluents, but receives no response from the state pollution Control Board then, how long should he wait to assume the 'unconditional consent'?
(A) 2 months (B) 4 months
(C) 6 months (D) 3 months

TEST 4 - KEYS

Q. Ans	Q. Ans	Q. Ans	Q. Ans	Q. Ans
1. C	2. B	3. D	4. A	5. B

TEST 5 : CLIMATE CHANGE

1. Dendroclimatology is the analysis of __________ to determine past climate variations

2. The Intergovernmental Panel on Climate Change (IPCC) was established by ________

3. ________can reduce the damaging effects of rising temperatures associated with climate change naturally

4. Factors that can shape climate are called ________

5. The study of contemporary and fossil palynomorphs, including pollen are called ________

6. FACE denotes
(A) Free Air Chemical Enrichment
(B) Free Air CO_2 Enrichment
(C) Free Air Carbon Enhancement
(D) Free Air Carbon Energy

7. Most of the warming observed over the past 50 years is attributable to
(A) Human activities (B) Volcanic eruption
(C) Forest fires (D) Continental drift

8. The anticipated change in global temperature of ________ or more degrees could be devastating, leading to significant maladaptations over the next rotation cycle of most tree species.
(A) 3.0 to 5.0 °C (B) 0.5 to 1.0 °C
(C) 0 to 0.5 °C (D) 1.0 to 1.5 °C

9. **Which among the following is not a component of sustainable forest framework**

   (A) Environment (B) Economic
   (C) Social (D) Political

10. **Which among the following is not considered as carbon sink**

    (A) Saline aquifer
    (B) Ocean water
    (C) Aging oil fields
    (D) Atmosphere

11. **________ is the capacity for a socio-ecological system to absorb stresses and maintain function in the face of external stresses imposed upon it by climate change**

    (A) Clima adopt (B) Climate resilience
    (C) Climate vulnerability (D) Climate forcing

12. **Climate system that responds more slowly to climate forcing are**

    (A) Ocean and ice caps (B) Mountain
    (C) Buildings (D) All

13. **Example for internal climate forcing mechanism**

    (A) Thermohaline circulations
    (B) Changes in solar output
    (C) Changes in CO_2 concentration
    (D) Changes in water vapour

14. **The eruption of Mount Pinatubo in 1991, the second largest terrestrial eruption of the 20th century, affected the climate substantially, subsequently global temperatures decreased by about**

    (A) 0.1 °C (B) 0.5 °C
    (C) 1.0 °C (D) 5.0 °C

15. **Which among the following is a clean technology to reduce emission?**

    (A) Carbon Fuel (B) Hydrogen fuel
    (C) Sulphur fuel (D) CNG

State TRUE or FALSE

16. **The expected sea level raise due to climate change is around 0.7 to 1.0 m.**

17. **All GHG are not measured in terms of CO_2.**

18. Trees play a key role in maintaining soil fertility because their leaves rot down and fertilize the soil.

19. Global warming is mainly because of the accumulation of Long wave length radiation.

20. Climate resilience efforts encompass Social, Economical, Political and Technological sectors.

TEST 6 : CLIMATE CHANGE

1. The fundamental part of climate system is ________
   (A) Soil (B) Ocean
   (C) Mountains (D) Volcanoes

2. The eruption of Mount Pinatubo in 1991, the second largest terrestrial eruption of the 20th century, affected the climate substantially, subsequently global temperatures decreased by about ________
   (A) 0.1 °C (B) 0.5 °C
   (C) 1.0 °C (D) 5.0 °C

3. The study of contemporary and fossil palynomorphs, including pollen is ________
   (A) Polynology (B) Pollenology
   (C) Palynology (D) Seismology

4. Dendroclimatology is the analysis of ________ to determine past climate variations
   (A) Tree pollen
   (B) Fossil palynomorph
   (C) tree ring growth patterns
   (D) ice core

5. the Intergovernmental Panel on Climate Change (IPCC) was established by ______
   (A) World Meteorological Organization (WMO)
   (B) World Health Organization (WHO)
   (C) Environmental Protection Agency (EPA)
   (D) Earth watch

6. Which among the following is a clean technology to reduce emission?
   (A) Carbon Fuel (B) Hydrogen fuel
   (C) Sulphur fuel (D) CNG

7. FACE denotes
   (A) Free Air Chemical Enrichment
   (B) Free Air CO_2 Enrichment
   (C) Free Air Carbon Enhancement
   (D) Free Air Carbon Energy

8. Most of the warming observed over the past 50 years is attributable to ______
   (A) Human activities (B) Volcanic eruption
   (C) Forest fires (D) Continental drift

9. The anticipated change in global temperature of ______ or more degrees could be devastating, leading to significant maladaptations over the next rotation cycle of most tree species.
   (A) 3.0 to 5.0 °C (B) 0.5 to 1.0 °C
   (C) 0 to 0.5 °C (D) 1.0 to 1.5 °C

10. Which among the following is not a component of sustainable forest framework ______
    (A) Environment (B) Economic
    (C) Social (D) Political

11. Factors that can shape climate are called ______
    (A) Climate forcing (B) Climate change
    (C) Climate drive (D) Weather drive

12. Climate system that responds more slowly to climate forcing are ______
    (A) Ocean and ice caps (B) Mountain
    (C) Buildings (D) All

13. Example for internal climate forcing mechanism ______
    (A) Thermohaline circulations
    (B) Changes in solar output
    (C) Changes in CO_2 concentration
    (D) Changes in water vapour

14. The response of the climate system will be fast, if ______
    (A) a sudden cooling due to airborne volcanic ash reflecting sunlight
    (B) thermal expansion of warming ocean water
    (C) sudden loss of albedo in the arctic ocean as sea ice melts
    (D) thermal expansion of the water

15. **Which among the following is not considered as carbon sink?**
(A) Saline aquifer (B) Ocean water
(C) Aging oil fields (D) Atmosphere

16. **Trees play a key role in maintaining soil fertility because ______**
(A) Their leaves rot down and fertilize the soil
(B) They capture the atmospheric Nitrogen
(C) They live for 100 years
(D) They absorbs more water

17. **______can reduce the damaging effects of rising temperatures associated with climate change naturally**
(A) Glaciers (B) Ocean
(C) Trees (D) Soil

18. **______ is the capacity for a socio-ecological system to absorb stresses and maintain function in the face of external stresses imposed upon it by climate change**
(A) Clima adopt (B) Climate resilience
(C) Climate vulnerability (D) Climate forcing

19. **Global warming is mainly because of the accumulation of ______**
(A) Short wave lengthy radiation
(B) Long wave length radiation
(C) Medium wave length radiation
(D) Both short and long wave length radiation

20. **Climate resilience efforts encompass ______**
(A) Social (B) Economical
(C) Political (D) All

21. **Factors that can shape climate are called ______**

22. **Climate system that responds more slowly to climate forcing are**

23. **Example for internal climate forcing mechanism IS ______**

24. **The eruption of Mount Pinatubo in 1991, the second largest terrestrial eruption of the 20th century, affected the climate substantially, subsequently global temperatures decreased by about ______**

25. **Dendroclimatology is the analysis of ______ to determine past climate variations**

26. Intergovernmental Panel on Climate Change (IPCC) was established by __________

27. The anticipated change in global temperature of ________ or more degrees could be devastating, leading to significant maladaptations over the next rotation cycle of most tree species.

28. ________can reduce the damaging effects of rising temperatures associated with climate change naturally

29. _________ is the capacity for a socio-ecological system to absorb stresses and maintain function in the face of external stresses imposed upon it by climate change

30. Climate resilience efforts encompass __________

TEST 7 : CLIMATE CHANGE

1. Factors that can shape climate are called ________

2. The study of contemporary and fossil palynomorphs, including pollen are called ________

3. Dendroclimatology is the analysis of __________ to determine past climate variations

4. The Intergovernmental Panel on Climate Change (IPCC) was established by ________

5. ______can reduce the damaging effects of rising temperatures associated with climate change naturally

6. Climate system that responds more slowly to climate forcing are
   (A) Ocean and ice caps (B) Mountain
   (C) Buildings (D) All

7. Example for internal climate forcing mechanism
   (A) Thermohaline circulations
   (B) Changes in solar output
   (C) Changes in CO_2 concentration
   (D) Changes in water vapour

8. The eruption of Mount Pinatubo in 1991, the second largest terrestrial eruption of the 20th century, affected the climate substantially, subsequently global temperatures decreased by about
   (A) 0.1 °C (B) 0.5 °C (C) 1.0 °C (D) 5.0 °C

9. Which among the following is a clean technology to reduce emission?
   (A) Carbon Fuel (B) Hydrogen fuel
   (C) Sulphur fuel (D) CNG

10. FACE denotes Free Air CO_2 Enrichment
   (A) Free Air Chemical Enrichment
   (B) Free Air CO_2 Enrichment
   (C) Free Air Carbon Enhancement
   (D) Free Air Carbon Energy

11. Most of the warming observed over the past 50 years is attributable to
   (A) Human activities (B) Volcanic eruption
   (C) Forest fires (D) Continental drift

12. The anticipated change in global temperature of ________ or more degrees could be devastating, leading to significant maladaptations over the next rotation cycle of most tree species.
   (A) 3.0 to 5.0 °C (B) 0.5 to 1.0 °C
   (C) 0 to 0.5 °C (D) 1.0 to 1.5 °C

13. Which among the following is not a component of sustainable forest framework
   (A) Environment (B) Economic
   (C) Social (D) Political

14. Which among the following is not considered as carbon sink
   (A) Saline aquifer (B) Ocean water
   (C) Aging oil fields (D) Atmosphere

15. ________ is the capacity for a socio-ecological system to absorb stresses and maintain function in the face of external stresses imposed upon it by climate change
   (A) Clima adopt (B) Climate resilience
   (C) Climate vulnerability (D) Climate forcing

TRUE or FALSE

16. Global warming is mainly because of the accumulation of Short wave length radiation

17. Climate resilience efforts encompass Social, Economical, Political and Technological sectors

18. **The response of the climate system will be fast, if a sudden cooling due to airborne volcanic ash reflecting sunlight**

19. **The fundamental part of climate system Soil**

20. **Trees play a key role in maintaining soil fertility because their leaves rot down and fertilize the soil**

◈

REFERENCE

Prasanthrajan, M and Mahendran, P.P. 2008. Text Book on Ecology and Environmental Science. ISBN 10: 8183211046, Agrotech Publishing Academy, Udaipur, India. pp. 1-285

Prasanthrajan, M and Balasubramanian, A. 2018. Objective Environmental Studies and Disaster Management. ISBN: 9789387893825, Scientific Publishers, India. pp. 1-146

Rajkishore, S.K. Davamani, V. Paul Sebastian, S. Sunitha, R. Prasanthrajan, M. and Sivasubramanian, K. 2019. Climate Change and Disaster Management, ISBN 9789387084162. A.E. Publications, Coimbatore, India. pp 230.

www.cpcb.com

www.currentaffairs.gktoday.in

www.epa.org

www.pakmcqs.com

www.slideshare.com

www.un.org

◈

Competitive Exam Preparation & Mcqs Books

Titles	*Authors*
— 9 year Questions: Sorted and solved CSIR UGC NET Life Science	*Kailash Chaudhary*
— Agricultural Extension A Glimpse	*S. Jananai*
— Agricultural Science at Your Fingertips**	*Jagan Singh Gora*
— Anatomy *(Vol.1-4)*: Exam Success Review MCQs for MBBS 1st Prof & PG Entrance	*Brijendra Singh*
— Anatomy for NEET PG Theory & MCQs *(Vol. 1-2)*	*Brijendra Singh*
— Basic Concepts of Plant Biotech. With MCQ	*Vijay Prakash*
— Basic Concepts of Plant Science	*S. K. Bangarwa*
— Biology for NEET *(English & Hindi)* (NCERT Based Objective) ** (set)	*Sumant Verma*
— Biology for NEET *(NCERT Based Objective)*	*Sumant Verma*
— Botany for NEET and other Medical Entrance	*Khwaja Salahuddin*
— Competitive Nematology	*N.G. Ravichandran*
— CSIR NET: Part A (Hindi)	*Christy Varghese*
— CSIR-NET General Aptitude: A New Outlook	*Christy Varghese*
— Dr. Singhs NEET PG MCQs for Dental Students	*Brijendra Singh*
— Dr. Singhs NEET PG MCQs for Medical Students	*Brijendra Singh*
— Forestry: A Subjective Guide for IFS Aspirants	*K. T. Parthiban*
— Fundamentals of Agriculture *2nd Ed. (Vol. 1-2) (Set)*	*R.L. Arya*
— Glimpse on General Agriculture	*P. Laxman Rao*
— Indiras Objective Agricultural Biotechnology *2nd Ed*	*R.L. Arya*
— Indiras Objective Agricultural Extension	*R.L. Arya*
— Indiras Objective Agriculture	*R.L. Arya*
— Indiras Objective Agronomy *2nd Ed*	*R.L. Arya*
— Indiras Objective Agronomy *3rd Ed*	*R.L. Arya*
— Jeev Vigyaan NEET *(NCERT Aadarit Vastunishth)*	*Sumant Verma*
— Key to Success in Agriculture	*B.N. Kalsariya*
— Krishi Jagat *2nd Edition*: Jet Evam Krishi Paryavekshak Pariksha Upyogi	*M.L. Jat*
— MCQ in Agricultural Microbiology	*D. G.Panpatte*
— MCQ on Soil and water conservation engineering**	*Ramesh Verma*
— MCQ' s in Plant Breeding Biotechnology and Seed Science	*K. Vanangamudi*
— MCQs in Microbiology: Advanced	*Balaram Mohapatra*

Titles	*Authors*
— Model Test Paper of Seed Science and Technology	*K.Vanangamudi*
— Multiple Choice Questions (MCQ) in Food Technology	*P.M. Kotecha*
— Objective Agribusiness Management *3rd Ed*	*S.R. Panigrahy*
— Objective Agribusiness Mgmt. *2nd Ed*	*S.R. Panigrahy*
— Objective Animal Physiology for JRF, SRF, ARS, NET, Civil Services & Other Competitive Examinations	*Arpita Mohapatra*
— Objective Environmental Studies and Disaster Management	*M. Prasanthrajan*
— Objective Food Science & Technology *3rd Ed*	*Deepak Mudgil*
— Objective Forestry: For All Compet. Examination *3rd Edition*	*K.T. Parthiban*
— Objective Home Science at a Glance	*Shanti Balda*
— Objective Life Science. MCQs for Life Science Examination *3rd Edition*	*Kailash Choudhary*
— Objective Plant Pathology *2nd Ed*	*Susanta Banik*
— Objective Plant Physiology *3rd Ed*	*P. Dwivedi*
— Objective Plant Science at a Glance	*D.K. Verma*
— Objective Seed Science and Technology *2nd Ed*	*K. Vanangamudi*
— Objective Type Questions and Answers in Veterinary Immunology	*T.R. Kannaki*
— Plant Pathology at a Glance (Encyclopedia of Plant Pathology) *2nd Ed* **	*D.P. Tripathi*
— Practice Paper and Mock Test for NEET: Botany	*Khwaja Salahuddin*
— Question Bank: Seed Science and Technology	*K. Vanangamudi*
— Questions and Answers in Environmental Science Practical	*R.K. Kamble*
— SET Life Science: Solved Exam Questions	*Kailash Choudhary*
— Universal Forestry *2nd Ed.* (UPSC, PCS, ARS/SRF /JRF/AFO, State PG & Ph.D. Entrance examinations and interviews of all Forest services)	*Mohit Husain*
— Universal Objective Forestry 2nd for IFS, JRF, SRF, NET, ACF, RFO, AFO, IBPS and Other Allied Exams	*Mohit Husain*
— Vedic Maths and Mental Arithmetic: A New Outlook **	*Christy Varghese*
— Vegetable Crops at a Glance	*S.K. Tyagi*

Text Book of Disaster Management

A.K. Shrivastava

India is prone to many natural and manmade disasters every year and thus possesses a physical and economical crisis for human being. If nothing is prepared it will difficult to deal with the many challenges following a disaster. Disaster management is a multidisciplinary area,covering a wide range of issues such as monitoring, forecasting, evacuation, search and rescue, relief, reconstruction and rehabilitation. The management aspects like activities, awareness and use of state of art technology can play an important role in the minimization of its impact on people and their properties. The foremost aim of writing this book is to empower the students by providing the basic and general knowledge, services to be rendered in a disaster situation. The framework of disaster management at national, state and district level is facilitates their role and responsibilities to be undertaken in disaster situation is described including the policy, planning, organizational, operational, and other matters.

The book covers all the task of implementation of the initiative of inculcating the culture of preparedness in the community as they are the first responders in case of a disaster. The book includes, what, how, when and by whom what should be done before, during and after a disaster takes place. The book highlights the impacts of disasters on life and economy and provides concise information for their management. The monitoring, forecasting of many natural disaster and role of ministry and departments are explained in this text book. How the rescue and rehabilitation to be undertaken to fulfill the day to day need of affected people is also provided. To comprehend the reading more lucid the book has been divided into different chapters. The book will be of interest to students, faculty, community leaders, self-trained disaster managers, social workers, media personnel, bureaucrats and citizen. It will be an asset to any student who are pursuing a degree or appearing for any competitive examination

CONTENTS

ISBN: 9789389412451 | Pages: 250 | Price ₹ : 345